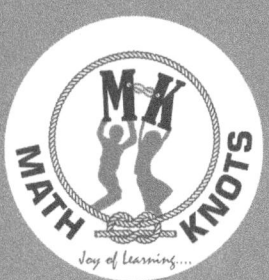

My Book

This book belongs to

Name: _____

Copy right © 2019 MATH-KNOTS LLC

All rights reserved, no part of this publication may be reproduced, stored in any system or transmitted in any form, or by any means, electronic, mechanical, photocopying, recording, or otherwise without the written permission of MATH-KNOTS LLC.

Cover Design by :
Gowri Vemuri

First Edition :
April, 2020

Second Edition : January, 2021

Author :
Gowri Vemuri

Editor :
Ritvik Pothapragada

Questions: mathknots.help@gmail.com

Dedication

This book is dedicated to:
My Mom, who is my best critic, guide and supporter.
To what I am today, and what I am going to become tomorrow,
is all because of your blessings, unconditional affection and support.

This book is dedicated to the
strongest women of my life,
my dearest mom
and
to all those moms in this universe.

G.V.

FRACTIONS

INDEX

Notes	9 - 22
Simplify fractions to lowest term	23 - 34
Simplify mixed fractions to lowest terms	35 - 42
Simplify improper fractions to lowest terms	43 - 52
Converting fractions to decimals	53 - 61
Fractions ; Addition	62 - 76
Fractions ; Subtraction	77 - 90
Fractions ; Addition and subtraction mixed	91 - 116
Fractions ; Multiplication	117 - 134
Fractions ; Division	135 - 142
Answer Keys	143 - 176

Fractions Notes

Fractions are part of a whole. It is also an expression representing quotient of two quantities.

Example 1 : $\frac{2}{4}, \frac{1}{4}$

Fraction can also be represented as ratios.

Example 2 : 1 : 2 or $\frac{1}{2}$

3 : 7 of $\frac{3}{7}$

Adding simple fractions, follow the below steps :

#1 : Fractions with like denominators can be added by adding their numerators.

Example 3 : $\frac{3}{5} + \frac{1}{5} = \frac{3+1}{5} = \frac{4}{5}$

#3 : To add fractions with unlike denominators, convert the fractions to equivalent fractions with like denominators and follow #1. To convert them into equivalent fractions you can multiply the numerator and denominator with a common factor for each of the fractions to be added separately and then add the fractions.

Method 1 :

Example 4 : $\frac{2}{5} + \frac{1}{2}$

Step 1 : $\frac{2}{5} = \frac{2 \times 2}{5 \times 2} = \frac{4}{10}$

Step 2 : $\frac{1}{2} = \frac{1 \times 5}{2 \times 5} = \frac{5}{10}$

Step 3 : $\frac{2}{5} + \frac{1}{2} = \frac{4}{10} + \frac{5}{10} = \frac{4+5}{10} = \frac{9}{10}$

$\frac{2}{5} + \frac{1}{2} = \frac{9}{10}$

 **FRACTIONS**

Another method to add fractions with unlike denominators is by finding the **Least Common Multiple** (LCM) of the denominators and then follow the steps as described below in same order.

Method 2 :

Example 5 : $\frac{5}{12} + \frac{1}{8}$

Step 1 : Find the LCM of the denominators 12 , 8

$$\begin{array}{r|l} 2 & 12,8 \\ 2 & 6,4 \\ 3 & 3,2 \\ 2 & 1,2 \\ & 1,1 \end{array}$$

LCM = 2 X 2 X 3 X 2 = 24

Step 2 : Converting $\frac{5}{12}$ into an equivalent fraction with the denominator as 24 (LCM).

Divide the LCM value with the number in the denominator of the fraction to obtain the common factor.

$$12 \overline{)\begin{array}{r} 2 \\ 24 \\ -24 \\ \hline 0 \end{array}}$$

Common factor obtained is 2

$\frac{5}{12} = \frac{5 \times 2}{12 \times 2} = \frac{10}{24}$; $\frac{5}{12} = \frac{10}{24}$

Step 3 : Converting $\frac{1}{8}$ into an equivalent fraction with the denominator as 24 (LCM).

Divide the LCM value with the number in the denominator of the fraction to obtain the common factor.

$$8 \overline{)\begin{array}{r} 3 \\ 24 \\ -24 \\ \hline 0 \end{array}}$$

Common factor obtained is 3

$\frac{1}{8} = \frac{1 \times 3}{8 \times 3} = \frac{3}{24}$; $\frac{1}{8} = \frac{3}{24}$

FRACTIONS

Step 4 :   Substitute the equivalent fractions obtained in step 2 and 3.

$$\frac{5}{12} + \frac{1}{8} = \frac{10}{24} + \frac{3}{24}$$

Step 5 :   The denominators of both fractions are same (Like denominators).
Add the numerators.

$$\frac{10+3}{24} = \frac{13}{24}$$

Note : To add more than two fractions, repeat step 2 or step 3 for each of the fractions to convert them into equivalent fractions and then proceed with step 4 and 5.

$$\frac{5}{12} + \frac{1}{8} = \frac{13}{24}$$

Subtracting simple fractions, follow the below steps :

#1 :   Fractions with like denominators can be subtracted by subtracting their numerators.

Example 6 : $\frac{3}{5} - \frac{1}{5} = \frac{3-1}{5} = \frac{2}{5}$

#2 :   To subtract fractions with unlike denominators, convert the fractions to equivalent fractions with like denominators and follow #1. To convert them into equivalent fractions you can multiply the numerator and denominator with a common factor for each of the fractions to be subtracted separately and then subtract the fractions.

Method 1 :

Example 7 : $\frac{3}{5} - \frac{1}{2}$

Step 1 : $\frac{3}{5} = \frac{3 \times 2}{5 \times 2} = \frac{6}{10}$

Step 2 : $\frac{1}{2} = \frac{1 \times 5}{2 \times 5} = \frac{5}{10}$

Step 3 : $\frac{3}{5} - \frac{1}{2} = \frac{6}{10} - \frac{5}{10} = \frac{6-5}{10} \quad \frac{1}{10}$

$$\frac{3}{5} - \frac{1}{2} = \frac{1}{10}$$

FRACTIONS

Notes

Another method to subtract fractions with unlike denominators is by finding the **Least Common Multiple** (LCM) of the denominators and then follow the steps as described below in same order.

Method 2 :

Example 8 : $\dfrac{5}{12} - \dfrac{1}{8}$

Step 1 :   Find the LCM of the denominators 12 , 8

```
2 | 12 , 8
2 |  6 , 4      LCM = 2 X 2 X 3 X 2 = 24
3 |  3 , 2
2 |  1 , 2
     1 , 1
```

Step 2 :   Converting $\dfrac{5}{12}$ into an equivalent fraction with the denominator as 24 (LCM).

Divide the LCM value with the number in the denominator of the fraction to obtain the common factor.

$$12 \overline{\smash{)}24} \;\; \substack{2 \\ -24 \\ \hline 0}$$

Common factor obtained is 2

$$\dfrac{5}{12} = \dfrac{5 \times 2}{12 \times 2} = \dfrac{10}{24} \;\; ; \;\; \dfrac{5}{12} = \dfrac{10}{24}$$

Step 3 :   Converting $\dfrac{1}{8}$ into an equivalent fraction with the denominator as 24 (LCM).

Divide the LCM value with the number in the denominator of the fraction to obtain the common factor.

$$8 \overline{\smash{)}24} \;\; \substack{3 \\ -24 \\ \hline 0}$$

Common factor obtained is 3

$$\dfrac{1}{8} \quad \dfrac{1 \times 3}{8 \times 3} \quad \dfrac{3}{24} \quad \dfrac{1}{8} \quad \dfrac{3}{24}$$

©All rights reserved-Math-Knots LLC., VA-USA
For more visit www.a4ace.com
www.math-knots.com

FRACTIONS

Notes

Step 4 :   Substitute the equivalent fractions obtained in step 2 and 3.

$$\frac{5}{12} - \frac{1}{8} = \frac{10}{24} - \frac{3}{24}$$

Step 5 :   The denominators of both fractions are same (Like denominators). Subtract the numerators.

$$\frac{10 - 3}{24} = \frac{7}{24}$$

Note : To Subtract more than two fractions, repeat step 2 or step 3 for each of the fractions to convert them into equivalent fractions and then proceed with step 4 and 5.

$$\frac{5}{12} - \frac{1}{8} = \frac{7}{24}$$

Adding Mixed numbers :

#1 :   Add the whole numbers together.

#2 :   Add the fractional parts together. (Find a common denominator if necessary) (Follow the steps as described in the previous pages)

#3 :   Write the whole number obtained from step 1 and the fraction obtained from step 2.

#4 :   If the fractional part is an improper fraction, change it to a mixed number. Add the whole part of the mixed number to the original whole numbers. Rewrite the fraction in the lowest possible value.

Example 9 : $3\frac{2}{7} + 5\frac{1}{7}$

Step 1 :   Add the whole part 3 from $3\frac{2}{7}$ to the whole part 5 from $5\frac{1}{7}$

$$3 + 5 = 8$$

Step 2 :   Add the fractional part $\frac{2}{7}$ from $3\frac{2}{7}$ to the fractional part $\frac{1}{7}$ from $5\frac{1}{7}$

$$\frac{2}{7} + \frac{1}{7} = \frac{2+1}{7} = \frac{3}{7}$$

$$3\frac{2}{7} + 5\frac{1}{7} = 8\frac{3}{7}$$

FRACTIONS

Notes

Example 10 : $4\frac{1}{6} + 5\frac{5}{6}$

Step 1 : Add the whole part 4 from $4\frac{1}{6}$ to the whole part 5 from $5\frac{5}{6}$

$4 + 5 = 9$

Step 2 : Add the fractional part $\frac{1}{6}$ from $4\frac{1}{6}$ to the fractional part $\frac{5}{6}$ from $5\frac{5}{6}$

$\frac{1}{6} + \frac{5}{6} = \frac{1+5}{6} = \frac{6}{6}$ (Numerators are added for fractions from like denominators)

Step 3 : $4\frac{1}{6} + 5\frac{5}{6} = 9\frac{6}{6} = 10$

($\frac{6}{6}$ = one whole part, add this one whole part to 9 making it equal to 10)

Example 11 : $7\frac{1}{2} + 5\frac{3}{20}$

Step 1 : Add the whole part 7 from $7\frac{1}{2}$ to the whole part 5 from $5\frac{3}{20}$

$7 + 5 = 12$

Step 2 : Add the fractional part $\frac{1}{2}$ from $7\frac{1}{2}$ to the fractional part $\frac{3}{20}$ from $5\frac{3}{20}$

$\frac{1}{2} + \frac{3}{20}$ (The fractions has unlike denominators)

Finding the LCM of 2 and 20

```
2 | 2 , 20
2 | 1 , 10     LCM of 2 and 20 = 2 X 2 X 5 = 20
5 | 1 , 5
    1 , 1
```

Step 3 : Let's make $\frac{1}{2}$ as an equivalent fraction with a denominator of 20.

$2\overline{)20}$ → 10, remainder 0   Common factor is 10

FRACTIONS

Notes

$$\frac{1}{2} = \frac{1 \times 10}{2 \times 10} = \frac{10}{20}$$

Step 4 : The fractional part $\frac{3}{20}$ of $5\frac{3}{20}$ has the same common denominator as 20.

We do not need to convert $\frac{3}{20}$ into another equivalent fraction.

Remember : The fractions can vary from problem to problem and students need to follow step 3 for all the fractional parts to convert them to equivalent fractions.

Step 5 : $\frac{1}{2} + \frac{3}{20} = \frac{10}{20} + \frac{3}{20} = \frac{10+3}{20} = \frac{13}{20}$

Step 6 : $7\frac{1}{2} + 5\frac{3}{20} = 8\frac{13}{20}$

Subtracting Mixed numbers :

#1 : Subtract the whole numbers together.

#2 : Subtract the fractional parts together. (Find a common denominator if necessary) (Follow the steps as described in the previous pages)

#3 : Write the whole number obtained from step 1 and the fraction obtained from step 2.

#4 : If the fractional part is an improper fraction, change it to a mixed number. Add the whole part of the mixed number to the whole number obtained in step 1. Rewrite the fraction in the lowest possible value.

Example 12 : $9\frac{2}{7} - 5\frac{1}{7}$

Step 1 : Subtract the whole part 5 from $5\frac{1}{7}$ from the whole part 9 from $3\frac{2}{7}$

$9 - 5 = 4$

Step 2 : Subtract the fractional part $\frac{1}{7}$ from $5\frac{1}{7}$ from the fractional part $\frac{2}{7}$ from $3\frac{2}{7}$

$\frac{2}{7} - \frac{1}{7} = \frac{2-1}{7} = \frac{1}{7}$

Step 3 : $9\frac{2}{7} - 5\frac{1}{7} = 4\frac{1}{7}$

 **FRACTIONS**

Notes

Example 13 : $7\frac{5}{6} - 5\frac{1}{6}$

Step 1 : Subtract the whole part 5 from $5\frac{1}{6}$ from the whole part 7 from $7\frac{1}{6}$

$7 - 5 = 2$

Step 2 : Subtract the fractional part $\frac{1}{6}$ from $5\frac{1}{6}$ to the fractional part $\frac{5}{6}$ from $7\frac{5}{6}$

$\frac{5}{6} - \frac{1}{6} = \frac{5-1}{6} = \frac{4}{6}$ (Numerators are added for fractions from like denominators)

Step 3 : $7\frac{5}{6} - 5\frac{1}{6} = 2\frac{4}{6} = 2\frac{2}{3}$

($\frac{4}{6} = \frac{2}{3}$ Equivalent fractions)

Example 14 : $7\frac{1}{2} - 5\frac{3}{20}$

Step 1 : Subtract the whole part 5 from $5\frac{3}{20}$ from the whole part 7 from $7\frac{1}{2}$

$7 - 5 = 2$

Step 2 : Subtract the fractional part $\frac{3}{20}$ from $5\frac{3}{20}$ to the fractional part $\frac{1}{2}$ from $7\frac{1}{2}$

$\frac{3}{20} - \frac{1}{2}$ (The fractions has unlike denominators)

Finding the LCM of 2 and 20

```
2 | 2 , 20
2 | 1 , 10     LCM of 2 and 20 = 2 X 2 X 5 = 20
5 | 1 , 5
    1 , 1
```

Step 3 : Let's make $\frac{1}{2}$ as an equivalent fraction with a denominator of 20.

```
     10
  2) 20      Common factor is 10
    -20
     ___
      0
```

FRACTIONS

$$\frac{1}{2} = \frac{1 \times 10}{2 \times 10} = \frac{10}{20}$$

Step 4: The fractional part $\frac{3}{20}$ of $5\frac{3}{20}$ has the same common denominator as 20.

We do not need to convert $\frac{3}{20}$ into another equivalent fraction.

Remember: The fractions can vary from problem to problem and students need to follow step 3 for all the fractional parts to convert them to equivalent fractions.

Step 5: $\frac{1}{2} - \frac{3}{20} = \frac{10}{20} - \frac{3}{20} = \frac{10-3}{20} = \frac{7}{20}$

Step 6: $7\frac{1}{2} - 5\frac{3}{20} = 2\frac{7}{20}$

Example 15: $5\frac{1}{7} - 3\frac{3}{7}$

Step 1: Subtract the whole part 3 from $3\frac{3}{7}$ from the whole part 5 from $5\frac{1}{7}$

$5 - 3 = 2$

Step 2: Subtract the fractional part $\frac{3}{7}$ from $3\frac{3}{7}$ from the fractional part $\frac{1}{7}$ from $5\frac{1}{7}$

We cannot subtract $\frac{3}{7}$ from $\frac{1}{7}$

So we need to rewrite the fraction $5\frac{1}{7}$

$5\frac{1}{7} = 4\frac{8}{7}$ (Remember in this fraction one whole part equals to seven. so when we take one whole part into the fraction form we need to add 7 to the value in the numerator which equals to 7 + 1 = 8)

Step 3: Repeat step 1

$5\frac{1}{7} - 3\frac{3}{7} = 4\frac{8}{7} - 3\frac{3}{7}$

Subtract the whole part 3 from $3\frac{3}{7}$ from the whole part 4 from $4\frac{8}{7}$

$4 - 3 = 1$

FRACTIONS

Notes

Step 4 :   Subtract the fractional part $\frac{3}{7}$ from $3\frac{3}{7}$ from the fractional part $\frac{8}{7}$ from $4\frac{8}{7}$

$$\frac{8}{7} - \frac{3}{7} = \frac{8-3}{7} = \frac{5}{7}$$

Step 5 : $5\frac{1}{7} - 3\frac{3}{7} = \frac{5}{7}$

Multiplying Fractions :

#1 :   Verify if the fractions are in lowest possible values. If not convert them into lowest possible values.

#2 :   Using cross simplification method simplify the fractions, meaning a numerator can be simplifies with a denominator only and vice versa.

#3 :   Do not cross simplify numerator with a numerator value and denominator with a denominator value

#4 :   Multiply the numerator with the remaining numerator values and the denominator with the denominator values

Remember : "Top times the top over the bottom times the bottom".
          All the answers must be written in simplest form.

Example 16 : $\frac{6}{15} \times \frac{3}{10}$

$$\frac{\overset{2}{\cancel{6}}}{\underset{5}{\cancel{15}}} \times \frac{3}{10} = \frac{\overset{1}{\cancel{2}}}{5} \times \frac{3}{\underset{5}{\cancel{10}}} = \frac{1}{5} \times \frac{3}{5}$$

$$= \frac{1 \times 3}{5 \times 5} = \frac{3}{25}$$

FRACTIONS

Notes

Multiplying Mixed Numbers :

#1 : To multiply mixed numbers, convert them to improper fractions.

Converting mixed number to improper fractions :
Multiply the denominator of the fraction to the whole part and then add the product to the numerator.

Example 17 : $2\frac{1}{3} = \frac{2 \times 3 + 1}{3} = \frac{6+1}{3} = \frac{7}{3}$

#2 : Verify if the fractions are in lowest possible values. If not convert them into lowest possible values.

#3 : Using cross simplification method simplify the fractions, meaning a numerator can be simplifies with a denominator only and vice versa.

#4 : Do not cross simplify numerator with a numerator value and denominator with a denominator value

#5 : Multiply the numerator with the remaining numerator values and the denominator with the denominator values

Remember : "Top times the top over the bottom times the bottom".
All the answers must be written in simplest form.
All improper fractions must be change back to mixed numbers.

Note : MULTIPLICATION CAN BE WRITTEN WITH THE SYMBOLS X OR . IN BETWEEN, .

FRACTIONS

Dividing Fractions :

#1 : To divide fractions, convert the division problem into a multiplication problem. Do this by multiplying the first fraction by the reciprocal of the second fraction. In other words convert the division to multiplication and interchange the numerator and denominator of the second fraction.

Remember : "When two fractions we divide, flip the second and multiply."

Note : Don't forget to check for "cross simplification" when multiplying.
All answers must be written in simplest form.
All improper fractions must be change back to mixed numbers.

Example 18 : $\dfrac{6}{15} \div \dfrac{3}{10}$

$$\dfrac{6}{15} \div \dfrac{3}{10} = \dfrac{\cancel{6}^2}{\cancel{15}_5} \times \dfrac{10}{3} = \dfrac{2}{\cancel{5}_1} \times \dfrac{\cancel{10}^2}{3} = \dfrac{2}{1} \times \dfrac{2}{3}$$

$$= \dfrac{2 \times 2}{1 \times 3} = \dfrac{4}{3}$$

Dividing Mixed Numbers :

#1 : To divide mixed fractions, first change them to improper fractions.

#2 : To divide fractions, convert the division problem into a multiplication problem. Do this by multiplying the first fraction by the reciprocal of the second fraction. In other words convert the division to multiplication and interchange the numerator and denominator of the second fraction.

Remember : "When two fractions we divide, flip the second and multiply."

Note : Don't forget to check for "cross simplification" when multiplying.
All answers must be written in simplest form.
All improper fractions must be change back to mixed numbers.

Example 19 : $2\dfrac{6}{15} \div 5\dfrac{3}{10}$

$$2\dfrac{6}{15} \div 5\dfrac{3}{10} = \dfrac{2 \times 15 + 6}{15} \div \dfrac{5 \times 10 + 3}{10} = \dfrac{30 + 6}{15} \div \dfrac{50 + 3}{10}$$

FRACTIONS

Notes

$= \dfrac{30 + 6}{15} \div \dfrac{50 + 3}{10}$

$= \dfrac{36}{15} \div \dfrac{53}{10}$

$= \dfrac{36}{15} \times \dfrac{10}{53}$ ← (Remember when we change division to multiplication, flip the second fraction)

$= \dfrac{\overset{12}{\cancel{36}}}{\underset{5}{\cancel{15}}} \times \dfrac{10}{53}$

$= \dfrac{12}{\underset{1}{\cancel{5}}} \times \dfrac{\overset{2}{\cancel{10}}}{53}$

$= \dfrac{12}{1} \times \dfrac{2}{53}$

$= \dfrac{12 \times 2}{1 \times 53}$

$= \dfrac{24}{53}$

$2\dfrac{6}{15} \div 5\dfrac{3}{10} = \dfrac{24}{53}$

FRACTIONS

Basic Math

Simplify each of the below fractions to the lowest possible numbers. Write your answer as a mixed number where possible.

(1) $\dfrac{4}{6}$

(2) $\dfrac{3}{9}$

(3) $\dfrac{12}{28}$

(4) $\dfrac{2}{8}$

(5) $\dfrac{12}{32}$

(6) $\dfrac{10}{15}$

(7) $\dfrac{2}{16}$

(8) $\dfrac{9}{21}$

(9) $\dfrac{63}{72}$

(10) $\dfrac{10}{16}$

FRACTIONS

Basic Math

Simplify each of the below fractions to the lowest possible numbers. Write your answer as a mixed number where possible.

(11) $\dfrac{8}{12}$ (12) $\dfrac{54}{63}$

(13) $\dfrac{10}{25}$ (14) $\dfrac{5}{30}$

(15) $\dfrac{27}{36}$ (16) $\dfrac{6}{24}$

(17) $\dfrac{6}{12}$ (18) $\dfrac{4}{12}$

(19) $\dfrac{20}{25}$ (20) $\dfrac{12}{42}$

 **FRACTIONS**

Basic Math

Simplify each of the below fractions to the lowest possible numbers. Write your answer as a mixed number where possible.

(21) $\dfrac{5}{15}$

(22) $\dfrac{5}{20}$

(23) $\dfrac{5}{10}$

(24) $\dfrac{6}{16}$

(25) $\dfrac{2}{4}$

(26) $\dfrac{8}{32}$

(27) $\dfrac{18}{30}$

(28) $\dfrac{24}{30}$

(29) $\dfrac{36}{60}$

(30) $\dfrac{4}{8}$

FRACTIONS

Basic Math

Simplify each of the below fractions to the lowest possible numbers. Write your answer as a mixed number where possible.

(31) $\dfrac{6}{54}$ \qquad (32) $\dfrac{140}{160}$

(33) $\dfrac{40}{64}$ \qquad (34) $\dfrac{9}{36}$

(35) $\dfrac{6}{42}$ \qquad (36) $\dfrac{18}{60}$

(37) $\dfrac{20}{80}$ \qquad (38) $\dfrac{12}{18}$

(39) $\dfrac{42}{60}$ \qquad (40) $\dfrac{36}{63}$

FRACTIONS

Basic Math

Simplify each of the below fractions to the lowest possible numbers. Write your answer as a mixed number where possible.

(41) $\dfrac{4}{12}$

(42) $\dfrac{18}{63}$

(43) $\dfrac{36}{81}$

(44) $\dfrac{54}{126}$

(45) $\dfrac{6}{18}$

(46) $\dfrac{20}{40}$

(47) $\dfrac{60}{96}$

(48) $\dfrac{80}{100}$

(49) $\dfrac{18}{36}$

(50) $\dfrac{18}{54}$

FRACTIONS

Basic Math

Simplify each of the below fractions to the lowest possible numbers. Write your answer as a mixed number where possible.

(51) $\dfrac{20}{52}$

(52) $\dfrac{9}{18}$

(53) $\dfrac{24}{90}$

(54) $\dfrac{12}{18}$

(55) $\dfrac{9}{135}$

(56) $\dfrac{18}{45}$

(57) $\dfrac{12}{78}$

(58) $\dfrac{198}{234}$

(59) $\dfrac{78}{90}$

(60) $\dfrac{198}{270}$

 **FRACTIONS**

Basic Math

Simplify each of the below fractions to the lowest possible numbers. Write your answer as a mixed number where possible.

(61) $\dfrac{45}{54}$ (62) $\dfrac{42}{90}$

(63) $\dfrac{63}{72}$ (64) $\dfrac{30}{36}$

(65) $\dfrac{45}{72}$ (66) $\dfrac{156}{168}$

(67) $\dfrac{12}{42}$ (68) $\dfrac{12}{90}$

(69) $\dfrac{156}{180}$ (70) $\dfrac{27}{45}$

FRACTIONS

Basic Math

Simplify each of the below fractions to the lowest possible numbers. Write your answer as a mixed number where possible.

(71) $\dfrac{90}{117}$ (72) $\dfrac{72}{156}$

(73) $\dfrac{28}{36}$ (74) $\dfrac{20}{40}$

(75) $\dfrac{36}{54}$ (76) $\dfrac{36}{126}$

(77) $\dfrac{100}{240}$ (78) $\dfrac{160}{200}$

(79) $\dfrac{280}{520}$ (80) $\dfrac{40}{140}$

 **FRACTIONS**

Basic Math

Simplify each of the below fractions to the lowest possible numbers. Write your answer as a mixed number where possible.

(81) $\dfrac{16}{64}$

(82) $\dfrac{120}{168}$

(83) $\dfrac{144}{324}$

(84) $\dfrac{108}{132}$

(85) $\dfrac{90}{234}$

(86) $\dfrac{140}{240}$

(87) $\dfrac{36}{540}$

(88) $\dfrac{24}{288}$

(89) $\dfrac{12}{156}$

(90) $\dfrac{72}{90}$

FRACTIONS

Basic Math

Simplify each of the below fractions to the lowest possible numbers. Write your answer as a mixed number where possible.

(91) $\dfrac{36}{108}$

(92) $\dfrac{40}{260}$

(93) $\dfrac{80}{220}$

(94) $\dfrac{396}{432}$

(95) $\dfrac{96}{180}$

(96) $\dfrac{20}{180}$

(97) $\dfrac{108}{120}$

(98) $\dfrac{24}{72}$

(99) $\dfrac{198}{252}$

(100) $\dfrac{200}{220}$

 **FRACTIONS**

Basic Math

Simplify each of the below fractions to the lowest possible numbers. Write your answer as a mixed number where possible.

(101) $8\dfrac{9}{18}$

(102) $3\dfrac{3}{6}$

(103) $7\dfrac{6}{18}$

(104) $2\dfrac{18}{27}$

(105) $7\dfrac{9}{15}$

(106) $1\dfrac{2}{4}$

(107) $6\dfrac{10}{15}$

(108) $4\dfrac{6}{10}$

(109) $5\dfrac{5}{25}$

(110) $1\dfrac{3}{18}$

FRACTIONS

Basic Math

Simplify each of the below fractions to the lowest possible numbers. Write your answer as a mixed number where possible.

(111) $5\dfrac{6}{8}$

(112) $8\dfrac{2}{12}$

(113) $4\dfrac{3}{12}$

(114) $4\dfrac{5}{25}$

(115) $4\dfrac{5}{35}$

(116) $4\dfrac{2}{4}$

(117) $7\dfrac{4}{6}$

(118) $2\dfrac{5}{20}$

(119) $1\dfrac{6}{15}$

(120) $4\dfrac{3}{21}$

 **FRACTIONS**

Basic Math

Simplify each of the below fractions to the lowest possible numbers. Write your answer as a mixed number where possible.

(121) $1\dfrac{5}{35}$

(122) $7\dfrac{6}{8}$

(123) $4\dfrac{12}{30}$

(124) $8\dfrac{2}{4}$

(125) $2\dfrac{5}{15}$

(126) $10\dfrac{9}{72}$

(127) $5\dfrac{6}{54}$

(128) $1\dfrac{24}{32}$

(129) $9\dfrac{12}{24}$

(130) $1\dfrac{9}{18}$

FRACTIONS

Basic Math

Simplify each of the below fractions to the lowest possible numbers. Write your answer as a mixed number where possible.

(131) $12\dfrac{18}{24}$ (132) $10\dfrac{24}{42}$

(133) $12\dfrac{9}{27}$ (134) $2\dfrac{18}{30}$

(135) $11\dfrac{20}{40}$ (136) $8\dfrac{6}{36}$

(137) $5\dfrac{4}{28}$ (138) $10\dfrac{54}{90}$

(139) $11\dfrac{8}{40}$ (140) $1\dfrac{36}{48}$

 **FRACTIONS**

Basic Math

Simplify each of the below fractions to the lowest possible numbers. Write your answer as a mixed number where possible.

(141) $2\dfrac{6}{24}$

(142) $1\dfrac{6}{48}$

(143) $7\dfrac{20}{100}$

(144) $3\dfrac{8}{32}$

(145) $8\dfrac{60}{100}$

(146) $3\dfrac{20}{60}$

(147) $9\dfrac{9}{18}$

(148) $9\dfrac{6}{12}$

(149) $7\dfrac{27}{63}$

(150) $8\dfrac{45}{54}$

FRACTIONS

Basic Math

Simplify each of the below fractions to the lowest possible numbers. Write your answer as a mixed number where possible.

(151) $4\dfrac{40}{60}$ (152) $10\dfrac{24}{120}$

(153) $4\dfrac{40}{120}$ (154) $8\dfrac{12}{60}$

(155) $6\dfrac{24}{120}$ (156) $3\dfrac{36}{72}$

(157) $10\dfrac{24}{60}$ (158) $12\dfrac{24}{108}$

(159) $11\dfrac{18}{54}$ (160) $3\dfrac{32}{40}$

 **FRACTIONS**

Basic Math

Simplify each of the below fractions to the lowest possible numbers. Write your answer as a mixed number where possible.

(161) $2\dfrac{20}{60}$

(162) $3\dfrac{12}{84}$

(163) $12\dfrac{40}{400}$

(164) $8\dfrac{24}{48}$

(165) $5\dfrac{120}{144}$

(166) $7\dfrac{36}{180}$

(167) $5\dfrac{80}{120}$

(168) $9\dfrac{48}{168}$

(169) $7\dfrac{18}{108}$

(170) $12\dfrac{80}{120}$

FRACTIONS

Basic Math

Simplify each of the below fractions to the lowest possible numbers. Write your answer as a mixed number where possible.

(171) $1\frac{24}{240}$

(172) $5\frac{8}{32}$

(173) $9\frac{18}{36}$

(174) $11\frac{12}{84}$

(175) $9\frac{24}{60}$

 **FRACTIONS**

Basic Math

Simplify each of the below fractions to the lowest possible numbers. Write your answer as an improper fraction.

(176) $\dfrac{54}{30}$

(177) $\dfrac{18}{8}$

(178) $\dfrac{27}{18}$

(179) $\dfrac{21}{12}$

(180) $\dfrac{12}{10}$

(181) $\dfrac{90}{81}$

(182) $\dfrac{42}{24}$

(183) $\dfrac{30}{12}$

(184) $\dfrac{14}{10}$

(185) $\dfrac{15}{12}$

FRACTIONS

Basic Math

Simplify each of the below fractions to the lowest possible numbers.
Write your answer as an improper fraction.

(186) $\dfrac{27}{15}$

(187) $\dfrac{6}{4}$

(188) $\dfrac{30}{18}$

(189) $\dfrac{63}{45}$

(190) $\dfrac{24}{15}$

(191) $\dfrac{27}{24}$

(192) $\dfrac{14}{6}$

(193) $\dfrac{21}{18}$

(194) $\dfrac{21}{9}$

(195) $\dfrac{10}{6}$

FRACTIONS

Basic Math

Simplify each of the below fractions to the lowest possible numbers. Write your answer as an improper fraction.

(196) $\dfrac{15}{6}$ (197) $\dfrac{8}{6}$

(198) $\dfrac{25}{20}$ (199) $\dfrac{30}{25}$

(200) $\dfrac{54}{48}$ (201) $\dfrac{342}{162}$

(202) $\dfrac{288}{54}$ (203) $\dfrac{72}{63}$

(204) $\dfrac{66}{42}$ (205) $\dfrac{78}{30}$

FRACTIONS

Basic Math

Simplify each of the below fractions to the lowest possible numbers. Write your answer as an improper fraction.

(206) $\dfrac{150}{108}$ (207) $\dfrac{42}{30}$

(208) $\dfrac{48}{30}$ (209) $\dfrac{36}{24}$

(210) $\dfrac{96}{76}$ (211) $\dfrac{72}{60}$

(212) $\dfrac{30}{12}$ (213) $\dfrac{153}{108}$

(214) $\dfrac{99}{36}$ (215) $\dfrac{68}{16}$

FRACTIONS

Basic Math

Simplify each of the below fractions to the lowest possible numbers. Write your answer as an improper fraction.

(216) $\dfrac{120}{100}$ (217) $\dfrac{72}{54}$

(218) $\dfrac{42}{24}$ (219) $\dfrac{264}{204}$

(220) $\dfrac{132}{114}$ (221) $\dfrac{189}{99}$

(222) $\dfrac{138}{108}$ (223) $\dfrac{102}{96}$

(224) $\dfrac{114}{48}$ (225) $\dfrac{460}{340}$

FRACTIONS

Basic Math

Simplify each of the below fractions to the lowest possible numbers. Write your answer as a mixed number.

(226) $\dfrac{288}{108}$

(227) $\dfrac{200}{180}$

(228) $\dfrac{100}{80}$

(229) $\dfrac{180}{72}$

(230) $\dfrac{72}{48}$

(231) $\dfrac{48}{36}$

(232) $\dfrac{324}{288}$

(233) $\dfrac{144}{64}$

(234) $\dfrac{180}{162}$

(235) $\dfrac{56}{32}$

 **FRACTIONS**

Basic Math

Simplify each of the below fractions to the lowest possible numbers. Write your answer as a mixed number.

(236) $\dfrac{200}{60}$

(237) $\dfrac{200}{160}$

(238) $\dfrac{108}{96}$

(239) $\dfrac{144}{108}$

(240) $\dfrac{240}{72}$

(241) $\dfrac{180}{144}$

(242) $\dfrac{80}{60}$

(243) $\dfrac{36}{24}$

(244) $\dfrac{126}{72}$

(245) $\dfrac{60}{40}$

# FRACTIONS

Basic Math

Simplify each of the below fractions to the lowest possible numbers. Write your answer as a mixed number.

(246) $\dfrac{100}{40}$

(247) $\dfrac{72}{54}$

(248) $\dfrac{90}{54}$

(249) $\dfrac{90}{36}$

(250) $\dfrac{200}{120}$

(251) $\dfrac{140}{100}$

(252) $\dfrac{216}{180}$

(253) $\dfrac{312}{216}$

(254) $\dfrac{84}{72}$

(255) $\dfrac{144}{54}$

FRACTIONS

Basic Math

Simplify each of the below fractions to the lowest possible numbers. Write your answer as a mixed number.

(256) $\dfrac{72}{60}$ (257) $\dfrac{200}{140}$

(258) $\dfrac{156}{144}$ (259) $\dfrac{162}{126}$

(260) $\dfrac{120}{84}$ (261) $\dfrac{200}{60}$

(262) $\dfrac{24}{16}$ (263) $\dfrac{192}{120}$

(264) $\dfrac{90}{54}$ (265) $\dfrac{40}{32}$

FRACTIONS

Basic Math

Simplify each of the below fractions to the lowest possible numbers. Write your answer as a mixed number.

(266) $\dfrac{80}{60}$ (267) $\dfrac{120}{100}$

(268) $\dfrac{200}{160}$ (269) $\dfrac{126}{90}$

(270) $\dfrac{260}{100}$ (271) $\dfrac{216}{90}$

(272) $\dfrac{104}{32}$ (273) $\dfrac{126}{54}$

(274) $\dfrac{60}{40}$ (275) $\dfrac{312}{168}$

FRACTIONS

Basic Math

Simplify each of the below fractions to the decimals.
Write the repeating decimals when necessary.

(276) $\dfrac{3}{5}$

(277) $\dfrac{3}{10}$

(278) $\dfrac{91}{100}$

(279) $6\dfrac{5}{33}$

(280) $4\dfrac{3}{4}$

(281) $\dfrac{5}{8}$

(282) $\dfrac{67}{100}$

(283) $\dfrac{7}{8}$

(284) $\dfrac{5}{11}$

(285) $9\dfrac{2}{3}$

FRACTIONS

Basic Math

Simplify each of the below fractions to the decimals.
Write the repeating decimals when necessary.

(286) $2\dfrac{1}{4}$

(287) $\dfrac{1}{10}$

(288) $\dfrac{1}{5}$

(289) $\dfrac{1}{2}$

(290) $3\dfrac{1}{2}$

(291) $\dfrac{1}{3}$

(292) $\dfrac{5}{33}$

(293) $\dfrac{5}{999}$

(294) $\dfrac{1}{8}$

(295) $\dfrac{9}{10}$

FRACTIONS

Simplify each of the below fractions to the decimals. Write the repeating decimals when necessary.

(296) $\dfrac{7}{10}$

(297) $6\dfrac{1}{2}$

(298) $\dfrac{2}{3}$

(299) $6\dfrac{94}{125}$

(300) $8\dfrac{22}{25}$

FRACTIONS

Basic Math

Simplify each of the below fractions to the decimals.
Round the decimals to the hundredths place.

(301) $\dfrac{7}{8}$

(302) $\dfrac{1}{180}$

(303) $\dfrac{20}{61}$

(304) $1\dfrac{14}{95}$

(305) $\dfrac{31}{100}$

(306) $5\dfrac{1}{2}$

(307) $8\dfrac{3}{25}$

(308) $1\dfrac{3}{5}$

(309) $1\dfrac{3}{10}$

(310) $\dfrac{1}{4}$

 **FRACTIONS**

Basic Math

Simplify each of the below fractions to the decimals.
Round the decimals to the hundredths place.

(311) $\dfrac{13}{31}$ (312) $\dfrac{1}{135}$

(313) $\dfrac{1}{248}$ (314) $3\dfrac{1}{2}$

(315) $5\dfrac{1}{8}$ (316) $\dfrac{1}{8}$

(317) $\dfrac{11}{30}$ (318) $\dfrac{1}{10}$

(319) $9\dfrac{1}{5}$ (320) $\dfrac{7}{10}$

FRACTIONS

Simplify each of the below fractions to the decimals.
Round the decimals to the hundredths place.

(321) $\dfrac{5}{8}$

(322) $\dfrac{1}{280}$

(323) $4\dfrac{41}{100}$

(324) $9\dfrac{12}{17}$

(325) $\dfrac{24}{35}$

 **FRACTIONS**

Basic Math

Simplify each of the below fractions to the decimals.
Round the decimals to the thousandths place.

(326) $\dfrac{21}{50}$

(327) $\dfrac{9}{10}$

(328) $\dfrac{1}{107}$

(329) $6\dfrac{3}{4}$

(330) $\dfrac{11}{21}$

(331) $\dfrac{5}{8}$

(332) $7\dfrac{3}{4}$

(333) $\dfrac{1}{100}$

(334) $4\dfrac{1}{2}$

(335) $6\dfrac{29}{40}$

FRACTIONS

Basic Math

Simplify each of the below fractions to the decimals.
Round the decimals to the thousandths place.

(336) $\dfrac{1}{130}$

(337) $\dfrac{1}{800}$

(338) $1\dfrac{53}{99}$

(339) $\dfrac{1}{280}$

(340) $\dfrac{49}{90}$

(341) $\dfrac{49}{65}$

(342) $6\dfrac{37}{50}$

(343) $\dfrac{1}{6}$

(344) $\dfrac{1}{220}$

(345) $\dfrac{1}{2}$

FRACTIONS

Basic Math

Simplify each of the below fractions to the decimals.
Round the decimals to the thousandths place.

(346) $\dfrac{1}{4}$

(347) $\dfrac{24}{25}$

(348) $\dfrac{7}{8}$

(349) $\dfrac{1}{5}$

(350) $\dfrac{7}{10}$

FRACTIONS

Basic Math

Simplify the below fractions by adding or subtracting as given below.

(351) $\dfrac{1}{2} + \dfrac{3}{2}$

(352) $\dfrac{1}{2} + \dfrac{1}{2}$

(353) $1 + 2\dfrac{1}{2}$

(354) $\dfrac{1}{2} + 4\dfrac{1}{2}$

(355) $\dfrac{5}{4} + \dfrac{7}{4}$

(356) $3\dfrac{1}{3} + 4\dfrac{2}{3}$

(357) $\dfrac{4}{7} + \dfrac{11}{7}$

(358) $1\dfrac{4}{5} + 4\dfrac{3}{5}$

(359) $\dfrac{5}{3} + \dfrac{5}{3}$

(360) $1 + \dfrac{5}{8}$

FRACTIONS

Basic Math

Simplify the below fractions by adding or subtracting as given below.

(361) $\dfrac{1}{2} + 6\dfrac{1}{2}$ (362) $2\dfrac{3}{4} + \dfrac{3}{4}$

(363) $\dfrac{5}{6} + \dfrac{7}{6}$ (364) $2\dfrac{3}{5} + 2\dfrac{2}{5}$

(365) $\dfrac{5}{6} + \dfrac{1}{6}$ (366) $\dfrac{3}{7} + \dfrac{2}{7}$

(367) $2\dfrac{1}{6} + \dfrac{5}{6}$ (368) $\dfrac{2}{3} + \dfrac{5}{3}$

(369) $\dfrac{5}{8} + 4\dfrac{5}{8}$ (370) $\dfrac{1}{2} + 2\dfrac{1}{2}$

 **FRACTIONS**

Simplify the below fractions by adding or subtracting as given below.

(371) $8 + \dfrac{4}{3}$

(372) $\dfrac{1}{8} + 2\dfrac{5}{8}$

(373) $4\dfrac{1}{4} + \dfrac{3}{4}$

(374) $\dfrac{11}{8} + 2\dfrac{1}{8}$

(375) $\dfrac{1}{3} + \dfrac{2}{3}$

(376) $1\dfrac{1}{6} + 4\dfrac{1}{9} + 5\dfrac{4}{5}$

(377) $\dfrac{5}{8} + 3\dfrac{1}{4} + 5\dfrac{5}{6}$

(378) $5\dfrac{3}{4} + 4\dfrac{7}{8} + 2$

(379) $\dfrac{7}{12} + \dfrac{12}{11} + 2$

(380) $6\dfrac{3}{5} + \dfrac{3}{10} + \dfrac{8}{11}$

FRACTIONS

Basic Math

Simplify the below fractions by adding or subtracting as given below.

(381) $\dfrac{9}{7} + 1\dfrac{1}{3} + 2$

(382) $2\dfrac{1}{7} + 4\dfrac{4}{7} + 1\dfrac{3}{8}$

(383) $10 + \dfrac{1}{2} + 5\dfrac{1}{2}$

(384) $\dfrac{11}{6} + \dfrac{4}{3} + \dfrac{7}{4}$

(385) $4\dfrac{5}{7} + \dfrac{7}{8} + \dfrac{4}{9}$

(386) $\dfrac{4}{3} + 1 + 2\dfrac{1}{9}$

(387) $\dfrac{1}{8} + \dfrac{5}{6} + 2$

(388) $\dfrac{8}{7} + 5\dfrac{1}{5} + 1\dfrac{5}{8}$

(389) $\dfrac{1}{2} + 3\dfrac{1}{12} + \dfrac{1}{12}$

(390) $6\dfrac{11}{12} + 3\dfrac{3}{8} + \dfrac{2}{3}$

FRACTIONS

Basic Math

Simplify the below fractions by adding or subtracting as given below.

(391) $\dfrac{4}{11} + 10\dfrac{1}{6} + \dfrac{7}{4}$

(392) $2 + 5\dfrac{7}{12} + \dfrac{5}{8}$

(393) $4\dfrac{7}{12} + 2 + \dfrac{6}{5}$

(394) $\dfrac{17}{9} + \dfrac{4}{3} + 1$

(395) $\dfrac{3}{2} + 2 + \dfrac{3}{2}$

(396) $\dfrac{2}{3} + \dfrac{4}{7} + 1\dfrac{2}{9}$

(397) $1\dfrac{4}{7} + \dfrac{1}{8} + 5$

(398) $\dfrac{11}{12} + \dfrac{1}{3} + \dfrac{1}{5}$

(399) $\dfrac{1}{11} + 1 + 1\dfrac{2}{3}$

(400) $5\dfrac{3}{7} + \dfrac{7}{5} + 1$

FRACTIONS

Basic Math

Simplify the below fractions by adding or subtracting as given below.

(401) $\dfrac{1}{7} + 5\dfrac{5}{7} + 6\dfrac{1}{7} + \dfrac{6}{7}$

(402) $\dfrac{2}{11} + \dfrac{20}{11} + \dfrac{14}{11} + 3\dfrac{4}{11}$

(403) $\dfrac{21}{13} + 10 + \dfrac{21}{13} + \dfrac{19}{13}$

(404) $\dfrac{6}{13} + 15\dfrac{9}{13} + \dfrac{15}{13} + 4\dfrac{7}{13}$

(405) $4 + \dfrac{5}{14} + \dfrac{9}{14} + 1\dfrac{3}{14}$

(406) $\dfrac{29}{15} + 4\dfrac{8}{15} + \dfrac{28}{15} + \dfrac{19}{15}$

(407) $2\dfrac{2}{5} + \dfrac{1}{5} + 2\dfrac{4}{5} + 4\dfrac{1}{5}$

(408) $3\dfrac{7}{11} + 0 + \dfrac{20}{11} + \dfrac{4}{11}$

(409) $\dfrac{11}{7} + 2\dfrac{2}{7} + \dfrac{2}{7} + \dfrac{5}{7}$

(410) $1\dfrac{3}{5} + \dfrac{1}{5} + 14\dfrac{3}{5} + \dfrac{1}{5}$

FRACTIONS

Basic Math

Simplify the below fractions by adding or subtracting as given below.

(411) $\dfrac{23}{14} + 7\dfrac{11}{14} + \dfrac{5}{14} + \dfrac{25}{14}$

(412) $\dfrac{19}{10} + 6\dfrac{7}{10} + 5\dfrac{3}{10} + \dfrac{11}{10}$

(413) $5\dfrac{4}{7} + \dfrac{13}{7} + \dfrac{13}{7} + \dfrac{5}{7}$

(414) $6\dfrac{8}{9} + 1\dfrac{1}{9} + \dfrac{16}{9} + 7\dfrac{7}{9}$

(415) $1\dfrac{1}{13} + 4\dfrac{4}{13} + \dfrac{15}{13} + 16$

(416) $6\dfrac{11}{16} + \dfrac{1}{16} + \dfrac{7}{16} + 8\dfrac{13}{16}$

(417) $\dfrac{13}{14} + \dfrac{27}{14} + 4\dfrac{13}{14} + \dfrac{25}{14}$

(418) $1\dfrac{1}{13} + \dfrac{22}{13} + 3\dfrac{7}{13} + \dfrac{21}{13}$

(419) $8\dfrac{13}{14} + \dfrac{13}{14} + \dfrac{23}{14} + 6\dfrac{11}{14}$

(420) $\dfrac{9}{13} + 7\dfrac{3}{13} + 2\dfrac{2}{13} + 5\dfrac{4}{13}$

FRACTIONS

Basic Math

Simplify the below fractions by adding or subtracting as given below.

(421) $\dfrac{7}{8} + \dfrac{3}{8} + 16 + \dfrac{13}{8}$

(422) $4\dfrac{15}{16} + \dfrac{19}{16} + 3\dfrac{11}{16} + \dfrac{19}{16}$

(423) $\dfrac{21}{16} + \dfrac{27}{16} + 3\dfrac{7}{16} + 6$

(424) $2\dfrac{9}{16} + 3\dfrac{5}{16} + \dfrac{27}{16} + 5\dfrac{15}{16}$

(425) $\dfrac{20}{11} + 11\dfrac{3}{11} + 13\dfrac{1}{11} + \dfrac{10}{11}$

(426) $\dfrac{4}{3} + \dfrac{4}{5}$

(427) $\dfrac{8}{5} + \dfrac{13}{7}$

(428) $\dfrac{1}{8} + \dfrac{5}{7}$

(429) $\dfrac{2}{5} + 1\dfrac{5}{8}$

(430) $2\dfrac{1}{5} + \dfrac{1}{2}$

 **FRACTIONS**

Simplify the below fractions by adding or subtracting as given below.

(431) $\dfrac{1}{6} + 3\dfrac{5}{8}$ (432) $\dfrac{10}{7} + \dfrac{1}{4}$

(433) $\dfrac{1}{4} + \dfrac{7}{8}$ (434) $\dfrac{1}{2} + \dfrac{3}{8}$

(435) $3\dfrac{1}{2} + \dfrac{1}{2}$ (436) $1 + \dfrac{1}{5}$

(437) $4\dfrac{1}{6} + 3\dfrac{1}{7}$ (438) $\dfrac{7}{4} + \dfrac{1}{4}$

(439) $4\dfrac{2}{3} + \dfrac{1}{7}$ (440) $\dfrac{10}{7} + \dfrac{5}{4}$

 **FRACTIONS**

Basic Math

Simplify the below fractions by adding or subtracting as given below.

(441) $\dfrac{7}{5} + \dfrac{5}{3}$

(442) $\dfrac{4}{7} + \dfrac{12}{7}$

(443) $4\dfrac{1}{2} + 3\dfrac{1}{6}$

(444) $\dfrac{3}{2} + 4\dfrac{4}{7}$

(445) $2\dfrac{1}{3} + 2\dfrac{1}{4}$

(446) $\dfrac{1}{8} + \dfrac{3}{2}$

(447) $1\dfrac{1}{2} + 4\dfrac{1}{6}$

(448) $3\dfrac{3}{4} + 3\dfrac{1}{6}$

(449) $1\dfrac{1}{4} + 1\dfrac{5}{8}$

(450) $\dfrac{2}{3} + \dfrac{2}{3}$

 **FRACTIONS**

Simplify the below fractions by adding or subtracting as given below.

(451) $5\dfrac{7}{10} + 2 + \dfrac{6}{5}$

(452) $5\dfrac{3}{7} + 8\dfrac{9}{10} + 8\dfrac{1}{3}$

(453) $\dfrac{7}{15} + 7\dfrac{7}{10} + \dfrac{5}{3}$

(454) $\dfrac{7}{16} + 11\dfrac{11}{14} + 4\dfrac{6}{7}$

(455) $6\dfrac{1}{3} + 2\dfrac{3}{8} + 6\dfrac{5}{16}$

(456) $\dfrac{11}{14} + 2 + \dfrac{5}{3}$

(457) $\dfrac{7}{6} + 3\dfrac{5}{12} + \dfrac{9}{16}$

(458) $2 + \dfrac{2}{7} + \dfrac{1}{8}$

(459) $\dfrac{9}{10} + \dfrac{19}{10} + \dfrac{1}{2}$

(460) $\dfrac{6}{7} + 1\dfrac{5}{9} + 3\dfrac{11}{15}$

 **FRACTIONS**

Basic Math

Simplify the below fractions by adding or subtracting as given below.

(461) $\dfrac{1}{4} + 2\dfrac{1}{15} + \dfrac{5}{4}$

(462) $1\dfrac{5}{8} + 1 + \dfrac{16}{15}$

(463) $\dfrac{6}{5} + \dfrac{1}{3} + 1\dfrac{5}{14}$

(464) $6\dfrac{1}{11} + 2 + \dfrac{6}{7}$

(465) $\dfrac{1}{3} + 8\dfrac{1}{2} + 1$

(466) $2\dfrac{3}{5} + \dfrac{5}{8} + \dfrac{3}{11}$

(467) $\dfrac{4}{3} + \dfrac{1}{3} + \dfrac{1}{9}$

(468) $3\dfrac{4}{5} + 5\dfrac{3}{10} + 8\dfrac{1}{4}$

(469) $3\dfrac{10}{13} + 2\dfrac{3}{8} + \dfrac{2}{3}$

(470) $\dfrac{2}{3} + 7\dfrac{1}{6} + 13$

 **FRACTIONS**

Basic Math

Simplify the below fractions by adding or subtracting as given below.

(471) $2 + \dfrac{21}{11} + 10$

(472) $8\dfrac{9}{10} + 3\dfrac{1}{8} + 4\dfrac{1}{3}$

(473) $6\dfrac{8}{11} + \dfrac{1}{2} + 7\dfrac{2}{7}$

(474) $2\dfrac{3}{8} + \dfrac{6}{7} + 6\dfrac{5}{16}$

(475) $\dfrac{1}{2} + 8 + \dfrac{7}{6}$

(476) $2\dfrac{5}{6} + \dfrac{4}{5} + \dfrac{10}{7} + 1$

(477) $\dfrac{1}{3} + 1 + \dfrac{1}{2} + \dfrac{7}{5}$

(478) $\dfrac{2}{7} + \dfrac{7}{4} + 1\dfrac{2}{5} + 1\dfrac{5}{6}$

(479) $\dfrac{5}{8} + 1 + \dfrac{3}{2} + \dfrac{5}{6}$

(480) $\dfrac{9}{7} + 1\dfrac{5}{6} + 2\dfrac{5}{8} + \dfrac{3}{4}$

FRACTIONS

Basic Math

Simplify the below fractions by adding or subtracting as given below.

(481) $\dfrac{1}{6} + \dfrac{1}{8} + 3\dfrac{3}{4} + \dfrac{5}{4}$

(482) $4 + \dfrac{5}{4} + 2 + \dfrac{3}{5}$

(483) $\dfrac{2}{5} + \dfrac{3}{7} + \dfrac{3}{5} + 1\dfrac{1}{2}$

(484) $\dfrac{1}{2} + 1\dfrac{1}{3} + \dfrac{2}{5} + 4\dfrac{3}{5}$

(485) $2 + 1\dfrac{1}{4} + 3\dfrac{2}{7} + 1\dfrac{1}{6}$

(486) $\dfrac{1}{2} + 7 + \dfrac{1}{4} + 2\dfrac{5}{8}$

(487) $3\dfrac{1}{2} + 3\dfrac{6}{7} + 6\dfrac{3}{4} + 1\dfrac{4}{5}$

(488) $\dfrac{4}{3} + 2 + 2\dfrac{1}{2} + 3\dfrac{1}{5}$

(489) $8 + \dfrac{5}{3} + 1\dfrac{1}{2} + \dfrac{1}{4}$

(490) $\dfrac{1}{2} + \dfrac{3}{2} + 1\dfrac{1}{3} + \dfrac{6}{7}$

FRACTIONS

Basic Math

Simplify the below fractions by adding or subtracting as given below.

(491) $3\dfrac{3}{4} + \dfrac{6}{5} + 0 + 4\dfrac{6}{7}$

(492) $4\dfrac{3}{7} + 1 + 3\dfrac{1}{6} + 3\dfrac{2}{7}$

(493) $3\dfrac{1}{8} + \dfrac{4}{3} + \dfrac{5}{4} + \dfrac{3}{2}$

(494) $7 + 1 + 2\dfrac{2}{5} + 3\dfrac{2}{3}$

(495) $\dfrac{1}{3} + \dfrac{4}{3} + 1\dfrac{5}{8} + \dfrac{1}{2}$

(496) $6 + \dfrac{2}{3} + \dfrac{1}{3} + 1\dfrac{7}{8}$

(497) $\dfrac{9}{7} + 4\dfrac{3}{4} + 3 + 1\dfrac{1}{5}$

(498) $\dfrac{11}{6} + 4\dfrac{1}{6} + \dfrac{4}{3} + \dfrac{1}{2}$

(499) $3\dfrac{2}{7} + 4\dfrac{1}{5} + 1 + \dfrac{1}{4}$

(500) $\dfrac{2}{3} + \dfrac{4}{3} + 3\dfrac{1}{4} + \dfrac{4}{3}$

FRACTIONS

Basic Math

Simplify the below fractions by adding or subtracting as given below.

(501) $1\dfrac{1}{3} - \dfrac{1}{3}$

(502) $3\dfrac{1}{6} - \dfrac{7}{6}$

(503) $1\dfrac{1}{7} - \dfrac{6}{7}$

(504) $2\dfrac{1}{6} - \dfrac{1}{6}$

(505) $4\dfrac{2}{3} - \dfrac{1}{3}$

(506) $4\dfrac{3}{8} - \dfrac{15}{8}$

(507) $3\dfrac{7}{8} - 3\dfrac{7}{8}$

(508) $\dfrac{7}{6} - \dfrac{1}{6}$

(509) $\dfrac{1}{2} - \dfrac{1}{2}$

(510) $4\dfrac{5}{6} - 4\dfrac{1}{6}$

FRACTIONS

Basic Math

Simplify the below fractions by adding or subtracting as given below.

(511) $\dfrac{10}{7} - \dfrac{4}{7}$

(512) $4\dfrac{1}{4} - \dfrac{5}{4}$

(513) $\dfrac{1}{4} - \dfrac{1}{4}$

(514) $2\dfrac{2}{5} - \dfrac{8}{5}$

(515) $\dfrac{15}{8} - 1\dfrac{5}{8}$

(516) $\dfrac{5}{4} - \dfrac{3}{4}$

(517) $\dfrac{1}{6} - \dfrac{1}{6}$

(518) $4\dfrac{1}{4} - \dfrac{3}{4}$

(519) $\dfrac{3}{2} - \dfrac{1}{2}$

(520) $1\dfrac{3}{5} - \dfrac{3}{5}$

FRACTIONS

Basic Math

Simplify the below fractions by adding or subtracting as given below.

(521) $2\dfrac{3}{4} - \dfrac{7}{4}$

(522) $3\dfrac{1}{2} - 3\dfrac{1}{2}$

(523) $2\dfrac{1}{2} - \dfrac{1}{2}$

(524) $2\dfrac{5}{6} - \dfrac{11}{6}$

(525) $3 - \dfrac{7}{8}$

(526) $\dfrac{24}{13} - \dfrac{10}{13} - \dfrac{10}{13}$

(527) $6\dfrac{13}{15} - \dfrac{14}{15} - \dfrac{7}{15}$

(528) $5\dfrac{1}{16} - 3\dfrac{15}{16} - \dfrac{7}{16}$

(529) $16 - 8\dfrac{1}{10} - 1\dfrac{1}{10}$

(530) $16 - 7\dfrac{1}{5} - 2\dfrac{2}{5}$

FRACTIONS

Basic Math

Simplify the below fractions by adding or subtracting as given below.

(531) $6\dfrac{5}{12} - \dfrac{11}{12} - \dfrac{5}{12}$

(532) $\dfrac{18}{13} - \dfrac{12}{13} - \dfrac{5}{13}$

(533) $8\dfrac{3}{10} - \dfrac{19}{10} - \dfrac{13}{10}$

(534) $5\dfrac{10}{13} - \dfrac{19}{13} - \dfrac{3}{13}$

(535) $3\dfrac{7}{8} - \dfrac{13}{8} - \dfrac{5}{8}$

(536) $1\dfrac{9}{14} - \dfrac{9}{14} - \dfrac{5}{14}$

(537) $7\dfrac{1}{5} - \dfrac{9}{5} - 2\dfrac{2}{5}$

(538) $16 - \dfrac{17}{16} - \dfrac{23}{16}$

(539) $8\dfrac{5}{9} - \dfrac{5}{9} - 3\dfrac{4}{9}$

(540) $7\dfrac{4}{7} - 1\dfrac{3}{7} - 3\dfrac{2}{7}$

FRACTIONS

Basic Math

Simplify the below fractions by adding or subtracting as given below.

(541) $\dfrac{15}{8} - \dfrac{11}{8} - \dfrac{3}{8}$

(542) $14\dfrac{4}{5} - 2\dfrac{4}{5} - \dfrac{3}{5}$

(543) $7\dfrac{3}{7} - \dfrac{13}{7} - 4\dfrac{6}{7}$

(544) $7\dfrac{6}{7} - 3\dfrac{6}{7} - \dfrac{3}{7}$

(545) $7\dfrac{5}{6} - \dfrac{1}{6} - \dfrac{7}{6}$

(546) $2\dfrac{7}{10} - \dfrac{11}{10} - \dfrac{9}{10}$

(547) $4\dfrac{11}{13} - \dfrac{10}{13} - 1\dfrac{5}{13}$

(548) $\dfrac{14}{9} - \dfrac{4}{9} - \dfrac{5}{9}$

(549) $9 - \dfrac{16}{9} - 3\dfrac{1}{9}$

(550) $7\dfrac{12}{13} - 1\dfrac{3}{13} - 3\dfrac{2}{13}$

FRACTIONS

Basic Math

Simplify the below fractions by adding or subtracting as given below.

(551) $3\dfrac{13}{28} - \dfrac{41}{28} - \dfrac{1}{28} - \dfrac{39}{28}$

(552) $23\dfrac{10}{23} - 4\dfrac{12}{23} - \dfrac{19}{23} - 16\dfrac{6}{23}$

(553) $14\dfrac{23}{30} - \dfrac{43}{30} - 2\dfrac{17}{30} - \dfrac{31}{30}$

(554) $46\dfrac{18}{19} - \dfrac{35}{19} - 21\dfrac{13}{19} - \dfrac{36}{19}$

(555) $48\dfrac{23}{37} - \dfrac{8}{37} - \dfrac{47}{37} - 12\dfrac{15}{37}$

(556) $22\dfrac{19}{23} - \dfrac{28}{23} - \dfrac{45}{23} - \dfrac{16}{23}$

(557) $15\dfrac{6}{19} - \dfrac{16}{19} - \dfrac{27}{19} - \dfrac{27}{19}$

(558) $23\dfrac{14}{45} - \dfrac{71}{45} - \dfrac{58}{45} - \dfrac{44}{45}$

(559) $17\dfrac{1}{43} - \dfrac{8}{43} - \dfrac{70}{43} - 11\dfrac{15}{43}$

(560) $29 - 19\dfrac{8}{39} - \dfrac{11}{39} - 2\dfrac{20}{39}$

FRACTIONS

Basic Math

Simplify the below fractions by adding or subtracting as given below.

(561) $44\dfrac{29}{49} - \dfrac{31}{49} - \dfrac{89}{49} - \dfrac{47}{49}$

(562) $37\dfrac{13}{23} - \dfrac{2}{23} - \dfrac{40}{23} - 23\dfrac{17}{23}$

(563) $42 - \dfrac{45}{29} - \dfrac{25}{29} - \dfrac{51}{29}$

(564) $17\dfrac{13}{18} - \dfrac{23}{18} - 13\dfrac{5}{18} - 1\dfrac{17}{18}$

(565) $27 - \dfrac{3}{16} - \dfrac{17}{16} - \dfrac{25}{16}$

(566) $8\dfrac{41}{49} - 4\dfrac{8}{49} - \dfrac{43}{49} - \dfrac{37}{49}$

(567) $4\dfrac{31}{41} - \dfrac{64}{41} - 2\dfrac{9}{41} - \dfrac{2}{41}$

(568) $10\dfrac{11}{20} - 7\dfrac{11}{20} - \dfrac{13}{20} - \dfrac{19}{20}$

(569) $8\dfrac{3}{46} - 2\dfrac{43}{46} - \dfrac{27}{46} - \dfrac{45}{46}$

(570) $5\dfrac{45}{49} - \dfrac{94}{49} - \dfrac{96}{49} - \dfrac{68}{49}$

FRACTIONS

Basic Math

Simplify the below fractions by adding or subtracting as given below.

(571) $14\dfrac{23}{27} - 1\dfrac{23}{27} - \dfrac{49}{27} - \dfrac{4}{27}$

(572) $8\dfrac{13}{29} - \dfrac{19}{29} - \dfrac{38}{29} - \dfrac{56}{29}$

(573) $8\dfrac{1}{33} - \dfrac{43}{33} - \dfrac{65}{33} - 2\dfrac{32}{33}$

(574) $19\dfrac{32}{35} - \dfrac{48}{35} - \dfrac{26}{35} - 6\dfrac{19}{35}$

(575) $24\dfrac{3}{34} - \dfrac{15}{34} - \dfrac{5}{34} - \dfrac{25}{34}$

(576) $4\dfrac{2}{5} - \dfrac{2}{3}$

(577) $4\dfrac{4}{5} - \dfrac{5}{8}$

(578) $2\dfrac{1}{4} - 1\dfrac{3}{8}$

(579) $1 - \dfrac{5}{6}$

(580) $\dfrac{7}{4} - 1\dfrac{3}{4}$

FRACTIONS

Basic Math

Simplify the below fractions by adding or subtracting as given below.

(581) $\dfrac{4}{5} - \dfrac{5}{7}$

(582) $\dfrac{6}{7} - \dfrac{1}{2}$

(583) $1\dfrac{5}{6} - \dfrac{3}{2}$

(584) $3\dfrac{7}{8} - \dfrac{5}{3}$

(585) $1\dfrac{1}{5} - \dfrac{1}{7}$

(586) $4 - 1\dfrac{3}{8}$

(587) $\dfrac{7}{5} - \dfrac{1}{7}$

(588) $1\dfrac{4}{5} - 1\dfrac{1}{3}$

(589) $\dfrac{3}{2} - \dfrac{1}{8}$

(590) $4\dfrac{3}{4} - \dfrac{1}{4}$

 **FRACTIONS**

Basic Math

Simplify the below fractions by adding or subtracting as given below.

(591) $3\dfrac{1}{8} - 2\dfrac{1}{2}$

(592) $4\dfrac{3}{4} - 3\dfrac{3}{7}$

(593) $4\dfrac{2}{3} - 1\dfrac{3}{5}$

(594) $2\dfrac{1}{2} - \dfrac{5}{6}$

(595) $4\dfrac{5}{6} - 4\dfrac{1}{2}$

(596) $5\dfrac{4}{9} - 2 - \dfrac{3}{2}$

(597) $\dfrac{3}{2} - \dfrac{3}{4} - \dfrac{3}{10}$

(598) $5\dfrac{2}{5} - \dfrac{1}{6} - \dfrac{5}{8}$

(599) $5\dfrac{7}{9} - 1 - \dfrac{1}{3}$

(600) $8 - \dfrac{1}{3} - 1\dfrac{1}{4}$

FRACTIONS

Basic Math

Simplify the below fractions by adding or subtracting as given below.

(601) $6\dfrac{3}{8} - \dfrac{3}{11} - \dfrac{5}{12}$

(602) $6\dfrac{1}{4} - 2 - \dfrac{5}{3}$

(603) $3 - \dfrac{5}{3} - \dfrac{13}{10}$

(604) $5\dfrac{1}{6} - 1 - 2\dfrac{7}{12}$

(605) $3\dfrac{7}{12} - \dfrac{1}{4} - 2\dfrac{10}{11}$

(606) $\dfrac{7}{4} - \dfrac{1}{5} - \dfrac{1}{2}$

(607) $3\dfrac{1}{2} - \dfrac{5}{8} - \dfrac{2}{7}$

(608) $1\dfrac{5}{6} - \dfrac{3}{5} - \dfrac{4}{9}$

(609) $3\dfrac{11}{12} - 2 - \dfrac{1}{3}$

(610) $10\dfrac{1}{4} - \dfrac{1}{8} - 6\dfrac{3}{4}$

FRACTIONS

Simplify the below fractions by adding or subtracting as given below.

(611) $6\dfrac{2}{5} - \dfrac{6}{5} - \dfrac{4}{5}$

(612) $11 - \dfrac{1}{3} - \dfrac{2}{7}$

(613) $6\dfrac{6}{7} - 3\dfrac{1}{2} - 2$

(614) $6\dfrac{1}{2} - \dfrac{1}{2} - \dfrac{1}{2}$

(615) $10 - \dfrac{8}{7} - \dfrac{17}{9}$

(616) $8\dfrac{1}{5} - 3\dfrac{1}{3} - \dfrac{3}{13} - 3\dfrac{5}{6}$

(617) $7\dfrac{5}{7} - 2\dfrac{11}{13} - \dfrac{5}{9} - 1\dfrac{6}{13}$

(618) $6\dfrac{8}{13} - 1\dfrac{1}{6} - \dfrac{7}{12} - \dfrac{1}{4}$

(619) $5\dfrac{1}{4} - 1\dfrac{5}{14} - 1 - \dfrac{8}{15}$

(620) $5 - 1\dfrac{3}{4} - \dfrac{1}{5} - \dfrac{1}{9}$

FRACTIONS

Basic Math

Simplify the below fractions by adding or subtracting as given below.

(621) $9\dfrac{4}{13} - \dfrac{5}{9} - 4\dfrac{5}{16} - 1$

(622) $5\dfrac{1}{2} - 3\dfrac{3}{4} - 1 - \dfrac{1}{2}$

(623) $8\dfrac{3}{4} - \dfrac{1}{15} - \dfrac{4}{7} - 3\dfrac{5}{6}$

(624) $4\dfrac{1}{3} - \dfrac{21}{16} - \dfrac{6}{7} - \dfrac{22}{13}$

(625) $9 - 1 - 2\dfrac{5}{12} - 1$

(626) $6\dfrac{5}{7} - \dfrac{15}{13} - 2 - \dfrac{12}{11}$

(627) $8\dfrac{4}{5} - \dfrac{1}{3} - \dfrac{7}{16} - 4\dfrac{1}{5}$

(628) $8 - \dfrac{2}{3} - \dfrac{3}{10} - 5\dfrac{12}{13}$

(629) $5\dfrac{4}{7} - \dfrac{5}{8} - \dfrac{5}{8} - 3\dfrac{1}{2}$

(630) $6\dfrac{3}{10} - 5\dfrac{1}{12} - \dfrac{3}{7} - \dfrac{5}{12}$

FRACTIONS

Simplify the below fractions by adding or subtracting as given below.

(631) $6\dfrac{3}{4} - 3\dfrac{11}{14} - \dfrac{14}{9} - \dfrac{9}{13}$

(632) $15\dfrac{5}{14} - \dfrac{1}{2} - 3\dfrac{2}{7} - \dfrac{8}{11}$

(633) $8\dfrac{10}{11} - \dfrac{1}{7} - \dfrac{5}{3} - 4\dfrac{5}{12}$

(634) $6\dfrac{1}{10} - \dfrac{15}{16} - \dfrac{10}{7} - \dfrac{1}{3}$

(635) $13\dfrac{11}{13} - 1\dfrac{2}{9} - 7\dfrac{7}{16} - \dfrac{1}{4}$

 **FRACTIONS**

Evaluate the below fractions to the lowest possible terms.

(636) $\dfrac{5}{3} + 4\dfrac{1}{2} + 3\dfrac{2}{3}$

(637) $\dfrac{5}{3} - \dfrac{3}{2} - \dfrac{1}{6}$

(638) $2\dfrac{1}{2} + 2 - \dfrac{1}{6}$

(639) $\dfrac{15}{8} + 2\dfrac{1}{4} - \dfrac{2}{7}$

(640) $2\dfrac{1}{6} + \dfrac{5}{3} + 1$

(641) $1\dfrac{1}{4} - \dfrac{2}{5} + 1\dfrac{3}{8}$

(642) $\dfrac{2}{3} + \dfrac{3}{2} + 3\dfrac{3}{7}$

(643) $\dfrac{1}{4} + 4\dfrac{1}{2} - 4\dfrac{3}{4}$

(644) $\dfrac{7}{4} + \dfrac{4}{7} - 1\dfrac{1}{4}$

(645) $1\dfrac{4}{5} + 6 - 2\dfrac{5}{8}$

FRACTIONS

Basic Math

Evaluate the below fractions to the lowest possible terms.

(646) $2 + 4\dfrac{3}{4} + \dfrac{5}{3}$

(647) $4 + \dfrac{2}{3} + \dfrac{5}{3}$

(648) $4\dfrac{1}{2} - 4\dfrac{3}{8} + 3\dfrac{6}{7}$

(649) $\dfrac{13}{7} + 2\dfrac{4}{5} + \dfrac{1}{8}$

(650) $\dfrac{5}{6} + 3\dfrac{1}{5} + \dfrac{2}{5}$

(651) $\dfrac{5}{8} + \dfrac{7}{4} + 1$

(652) $3\dfrac{2}{7} + \dfrac{3}{2} + \dfrac{2}{3}$

(653) $\dfrac{4}{5} + 1\dfrac{1}{4} + 3\dfrac{1}{4}$

(654) $\dfrac{9}{5} - 1 + 4\dfrac{1}{2}$

(655) $2\dfrac{5}{8} - \dfrac{1}{6} + 2\dfrac{1}{8}$

FRACTIONS

Basic Math

Evaluate the below fractions to the lowest possible terms.

(656) $\dfrac{5}{3} + \dfrac{5}{7} + 3\dfrac{7}{8}$

(657) $2\dfrac{1}{2} - 1 - \dfrac{1}{2}$

(658) $\dfrac{5}{3} - 1 + 2\dfrac{1}{4}$

(659) $2 + 1 - \dfrac{5}{3}$

(660) $3\dfrac{1}{7} - 2\dfrac{3}{5} + 1$

(661) $\dfrac{3}{2} + 3\dfrac{1}{7} - 4\dfrac{1}{6}$

(662) $4 + \dfrac{3}{2} - \dfrac{1}{2}$

(663) $1\dfrac{2}{3} + 4\dfrac{4}{7} - 2$

(664) $\dfrac{5}{4} + \dfrac{2}{3} + \dfrac{1}{3}$

(665) $\dfrac{11}{8} + 3\dfrac{1}{3} - \dfrac{11}{7}$

 **FRACTIONS**

Basic Math

Evaluate the below fractions to the lowest possible terms.

(666) $4\frac{5}{8} + 2\frac{7}{8}$

(667) $5\frac{1}{2} + \frac{1}{2}$

(668) $1\frac{1}{8} + \frac{7}{8}$

(669) $4\frac{1}{2} + \frac{1}{2}$

(670) $3\frac{1}{5} + \frac{2}{5}$

(671) $\frac{2}{3} + \frac{1}{3}$

(672) $2\frac{7}{8} + 1\frac{1}{8}$

(673) $1\frac{2}{7} + 4\frac{3}{7}$

(674) $1\frac{5}{6} + 2\frac{1}{6}$

(675) $8 + 4\frac{1}{2}$

 **FRACTIONS**

Basic Math

Evaluate the below fractions to the lowest possible terms.

(676) $2\dfrac{1}{3} + 2\dfrac{1}{3}$

(677) $4\dfrac{1}{3} + 1\dfrac{1}{3}$

(678) $3\dfrac{3}{5} + 1\dfrac{1}{5}$

(679) $7\dfrac{1}{4} + 3\dfrac{1}{4}$

(680) $4\dfrac{7}{8} + 4\dfrac{7}{8}$

(681) $3\dfrac{1}{2} + \dfrac{1}{2}$

(682) $\dfrac{1}{3} + 2\dfrac{2}{3}$

(683) $3\dfrac{2}{3} + 4\dfrac{1}{3}$

(684) $2\dfrac{1}{2} + 2\dfrac{1}{2}$

(685) $4\dfrac{7}{8} + 1\dfrac{5}{8}$

 **FRACTIONS**

Evaluate the below fractions to the lowest possible terms.

(686) $4\dfrac{3}{5} + \dfrac{2}{5}$

(687) $1\dfrac{3}{7} + 3\dfrac{5}{7}$

(688) $3\dfrac{5}{7} + 3\dfrac{2}{7}$

(689) $2\dfrac{1}{3} + 2\dfrac{2}{3}$

(690) $3\dfrac{4}{7} + 3\dfrac{6}{7}$

(691) $5\dfrac{1}{6} + 4\dfrac{5}{6} + 8\dfrac{1}{6}$

(692) $12\dfrac{3}{5} + 5\dfrac{1}{5} + 5\dfrac{1}{5}$

(693) $7\dfrac{2}{11} + 8\dfrac{9}{11} + 4\dfrac{3}{11}$

(694) $5\dfrac{4}{11} + 2\dfrac{2}{11} + 8\dfrac{2}{11}$

(695) $3\dfrac{1}{16} + 8\dfrac{15}{16} + 4\dfrac{9}{16}$

FRACTIONS

Basic Math

Evaluate the below fractions to the lowest possible terms.

(696) $6\frac{13}{15} + 6\frac{13}{15} + 2\frac{1}{15}$

(697) $8\frac{7}{12} + \frac{1}{12} + 8\frac{5}{12}$

(698) $3\frac{4}{5} + 8\frac{4}{5} + 8\frac{2}{5}$

(699) $2\frac{9}{10} + 6\frac{3}{10} + \frac{3}{10}$

(700) $1\frac{3}{5} + 2\frac{4}{5} + 9$

(701) $2\frac{1}{14} + 7\frac{9}{14} + 3\frac{13}{14}$

(702) $3\frac{5}{6} + 11 + \frac{5}{6}$

(703) $7\frac{8}{9} + 7 + 5\frac{5}{9}$

(704) $12\frac{11}{12} + \frac{5}{12} + \frac{7}{12}$

(705) $\frac{1}{5} + 6\frac{4}{5} + 6\frac{1}{5}$

FRACTIONS

Basic Math

Evaluate the below fractions to the lowest possible terms.

(706) $6\frac{1}{10} + 3\frac{7}{10} + 7\frac{1}{10}$

(707) $2\frac{1}{5} + 2\frac{2}{5} + 4\frac{2}{5}$

(708) $7\frac{13}{16} + 7\frac{3}{16} + 13$

(709) $8\frac{9}{13} + 1\frac{8}{13} + 7\frac{1}{13}$

(710) $\frac{13}{15} + 8\frac{8}{15} + 1\frac{13}{15}$

(711) $14 + 6\frac{2}{5} + 4\frac{3}{5}$

(712) $7\frac{10}{13} + 3\frac{8}{13} + 6\frac{1}{13}$

(713) $4\frac{2}{11} + 8\frac{7}{11} + 6\frac{2}{11}$

(714) $8\frac{1}{10} + 4\frac{1}{10} + 4\frac{1}{10}$

(715) $5\frac{5}{6} + 2\frac{5}{6} + \frac{5}{6}$

 **FRACTIONS**

Evaluate the below fractions to the lowest possible terms.

(716) $2\frac{7}{8} + \frac{1}{6}$

(717) $\frac{1}{7} + 4\frac{3}{4}$

(718) $1\frac{4}{7} + 3\frac{1}{6}$

(719) $1\frac{5}{6} + 1\frac{5}{8}$

(720) $2\frac{2}{3} + \frac{1}{2}$

(721) $2\frac{5}{8} + 7\frac{1}{5}$

(722) $3\frac{1}{8} + 2\frac{3}{5}$

(723) $8\frac{7}{8} + 2\frac{5}{7}$

(724) $2\frac{3}{8} + 4\frac{3}{8}$

(725) $1\frac{2}{7} + \frac{3}{8}$

 **FRACTIONS**

Basic Math

Evaluate the below fractions to the lowest possible terms.

(726) $2\frac{2}{5} + 1\frac{1}{5}$

(727) $3\frac{3}{4} + \frac{7}{8}$

(728) $1\frac{5}{8} + 4\frac{1}{4}$

(729) $1\frac{5}{8} + 3\frac{5}{8}$

(730) $\frac{1}{5} + 2\frac{3}{8}$

(731) $8\frac{3}{5} + \frac{1}{4}$

(732) $3\frac{2}{5} + 1\frac{5}{8}$

(733) $3\frac{3}{8} + 3\frac{1}{2}$

(734) $3\frac{2}{3} + \frac{1}{4}$

(735) $4\frac{1}{6} + 3\frac{1}{6}$

FRACTIONS

Basic Math

Evaluate the below fractions to the lowest possible terms.

(736) $3\dfrac{3}{5} + 4\dfrac{3}{8}$

(737) $4\dfrac{5}{8} + 4\dfrac{1}{4}$

(738) $1\dfrac{4}{5} + 4\dfrac{3}{8}$

(739) $2 + \dfrac{5}{7}$

(740) $4\dfrac{5}{6} + 4\dfrac{1}{3}$

(741) $4\dfrac{5}{8} + 11 + 3\dfrac{3}{4} + 12\dfrac{3}{11}$

(742) $2\dfrac{1}{8} + 5\dfrac{6}{7} + 4\dfrac{1}{2} + 4\dfrac{1}{12}$

(743) $3\dfrac{3}{4} + 1\dfrac{1}{3} + 6\dfrac{9}{10} + 6\dfrac{9}{10}$

(744) $6\dfrac{1}{12} + 8\dfrac{1}{5} + 1 + 4\dfrac{3}{4}$

(745) $1\dfrac{6}{11} + 4\dfrac{1}{12} + 6\dfrac{3}{4} + 5\dfrac{1}{2}$

 **FRACTIONS**

Basic Math

Evaluate the below fractions to the lowest possible terms.

(746) $4\dfrac{11}{12} + \dfrac{6}{11} + 5\dfrac{1}{4} + 4\dfrac{7}{9}$

(747) $1\dfrac{1}{2} + 1\dfrac{1}{9} + 1\dfrac{2}{3} + 6\dfrac{3}{4}$

(748) $6\dfrac{11}{12} + 5\dfrac{1}{2} + 2\dfrac{4}{7} + 5\dfrac{3}{7}$

(749) $\dfrac{3}{4} + 5\dfrac{9}{10} + 6\dfrac{3}{7} + 3\dfrac{1}{3}$

(750) $1\dfrac{9}{10} + 4\dfrac{7}{11} + \dfrac{2}{9} + \dfrac{4}{5}$

(751) $1 + \dfrac{6}{7} + 12 + 7$

(752) $4\dfrac{8}{9} + 4\dfrac{1}{2} + 4\dfrac{5}{9} + 5\dfrac{3}{7}$

(753) $\dfrac{1}{9} + 6\dfrac{5}{6} + 8 + 3\dfrac{7}{9}$

(754) $1\dfrac{5}{6} + 5\dfrac{8}{9} + 1\dfrac{1}{6} + 3\dfrac{2}{3}$

(755) $1\dfrac{7}{10} + 1 + 3\dfrac{3}{8} + 3\dfrac{3}{8}$

FRACTIONS

Basic Math

Evaluate the below fractions to the lowest possible terms.

(756) $6\dfrac{11}{12} + \dfrac{3}{4} + 3\dfrac{1}{2} + 3\dfrac{9}{10}$

(757) $10 + 4\dfrac{5}{6} + 3\dfrac{7}{12} + \dfrac{9}{10}$

(758) $3\dfrac{1}{2} + \dfrac{4}{5} + 10 + 5\dfrac{3}{8}$

(759) $\dfrac{11}{12} + \dfrac{1}{7} + 6\dfrac{4}{5} + \dfrac{3}{4}$

(760) $4\dfrac{3}{10} + 7 + 11 + 3\dfrac{3}{4}$

(761) $2\dfrac{2}{3} + 3\dfrac{10}{11} + 1\dfrac{5}{6} + 1\dfrac{1}{2}$

(762) $\dfrac{7}{9} + 1\dfrac{1}{2} + 1 + 5\dfrac{6}{11}$

(763) $5 + \dfrac{7}{8} + 2\dfrac{1}{4} + \dfrac{1}{12}$

(764) $5\dfrac{3}{4} + 2\dfrac{1}{2} + 3\dfrac{1}{2} + 3\dfrac{1}{2}$

(765) $6\dfrac{1}{5} + 7\dfrac{1}{2} + \dfrac{11}{12} + 2\dfrac{1}{3}$

 **FRACTIONS**

Basic Math

Evaluate the below fractions to the lowest possible terms.

(766) $4\frac{2}{3} - \frac{1}{3}$

(767) $3\frac{1}{2} - \frac{1}{2}$

(768) $4\frac{5}{8} - 1\frac{3}{8}$

(769) $6 - 4\frac{1}{4}$

(770) $3\frac{5}{6} - \frac{5}{6}$

(771) $4\frac{5}{6} - 1\frac{1}{6}$

(772) $8\frac{3}{5} - 3\frac{1}{5}$

(773) $3\frac{3}{8} - 2\frac{1}{8}$

(774) $4\frac{3}{4} - 1\frac{3}{4}$

(775) $\frac{1}{6} - \frac{1}{6}$

 **FRACTIONS**

Basic Math

Evaluate the below fractions to the lowest possible terms.

(776) $4\dfrac{1}{4} - 1\dfrac{1}{4}$

(777) $3\dfrac{2}{3} - 2\dfrac{2}{3}$

(778) $4\dfrac{1}{7} - 1\dfrac{4}{7}$

(779) $4\dfrac{1}{6} - 1\dfrac{5}{6}$

(780) $4\dfrac{1}{2} - 4\dfrac{1}{2}$

(781) $4\dfrac{5}{6} - 3\dfrac{1}{6}$

(782) $3\dfrac{2}{3} - \dfrac{1}{3}$

(783) $\dfrac{1}{2} - \dfrac{1}{2}$

(784) $6 - 3\dfrac{1}{3}$

(785) $3\dfrac{1}{2} - 3\dfrac{1}{2}$

FRACTIONS

Basic Math

Evaluate the below fractions to the lowest possible terms.

(786) $3\dfrac{1}{5} - \dfrac{1}{5}$

(787) $4\dfrac{5}{8} - 1\dfrac{1}{8}$

(788) $4\dfrac{3}{4} - 3\dfrac{1}{4}$

(789) $4\dfrac{6}{7} - \dfrac{5}{7}$

(790) $3\dfrac{5}{6} - 1\dfrac{1}{6}$

(791) $3\dfrac{1}{2} - 2\dfrac{1}{2} - \dfrac{1}{2}$

(792) $3\dfrac{3}{8} - 3 - \dfrac{3}{8}$

(793) $3\dfrac{1}{3} - 1\dfrac{1}{3} - 1\dfrac{2}{3}$

(794) $3\dfrac{1}{4} - 1\dfrac{1}{4} - 1\dfrac{3}{4}$

(795) $\dfrac{3}{4} - \dfrac{1}{4} - \dfrac{1}{4}$

FRACTIONS

Basic Math

Evaluate the below fractions to the lowest possible terms.

(796) $\quad 4\dfrac{6}{7} - 1\dfrac{6}{7} - 1\dfrac{3}{7}$

(797) $\quad 8\dfrac{1}{2} - \dfrac{1}{2} - 3\dfrac{1}{2}$

(798) $\quad 4\dfrac{2}{3} - \dfrac{2}{3} - \dfrac{2}{3}$

(799) $\quad 4\dfrac{4}{5} - \dfrac{3}{5} - 1\dfrac{1}{5}$

(800) $\quad 7 - \dfrac{3}{7} - 3\dfrac{4}{7}$

(801) $\quad 4\dfrac{3}{4} - \dfrac{1}{4} - 4\dfrac{1}{4}$

(802) $\quad 2\dfrac{7}{8} - 2\dfrac{3}{8} - \dfrac{3}{8}$

(803) $\quad 3\dfrac{7}{8} - 1\dfrac{1}{8} - 2\dfrac{5}{8}$

(804) $\quad 2\dfrac{4}{7} - \dfrac{4}{7} - 1\dfrac{5}{7}$

(805) $\quad 3\dfrac{1}{2} - 1\dfrac{1}{2} - \dfrac{1}{2}$

FRACTIONS

Evaluate the below fractions to the lowest possible terms.

(806) $\quad 4\dfrac{5}{7} - 2\dfrac{6}{7} - \dfrac{2}{7}$

(807) $\quad 4\dfrac{3}{8} - 1\dfrac{5}{8} - \dfrac{3}{8}$

(808) $\quad 4\dfrac{5}{6} - \dfrac{5}{6} - 2\dfrac{1}{6}$

(809) $\quad 2\dfrac{7}{8} - \dfrac{5}{8} - \dfrac{5}{8}$

(810) $\quad 3\dfrac{5}{8} - 1\dfrac{1}{8} - 1\dfrac{3}{8}$

(811) $\quad 4\dfrac{6}{7} - 1\dfrac{4}{7} - 1\dfrac{6}{7}$

(812) $\quad 3\dfrac{5}{6} - \dfrac{1}{6} - 2\dfrac{1}{6}$

(813) $\quad 4\dfrac{3}{4} - 1\dfrac{1}{4} - 3\dfrac{1}{4}$

(814) $\quad 4\dfrac{1}{4} - 2\dfrac{3}{4} - 1\dfrac{1}{4}$

(815) $\quad 2\dfrac{5}{7} - \dfrac{2}{7} - 2\dfrac{3}{7}$

FRACTIONS

Basic Math

Evaluate the below fractions to the lowest possible terms.

(816) $\quad 3\dfrac{1}{4} - \dfrac{4}{7}$

(817) $\quad 4\dfrac{1}{3} - 2\dfrac{3}{5}$

(818) $\quad 4\dfrac{1}{6} - \dfrac{7}{8}$

(819) $\quad 2\dfrac{4}{7} - 1\dfrac{6}{7}$

(820) $\quad 3\dfrac{2}{7} - 3\dfrac{1}{7}$

(821) $\quad 5\dfrac{4}{5} - 1\dfrac{1}{2}$

(822) $\quad 3\dfrac{5}{6} - \dfrac{5}{8}$

(823) $\quad 4\dfrac{1}{2} - 3\dfrac{1}{6}$

(824) $\quad 3\dfrac{2}{3} - 2\dfrac{4}{5}$

(825) $\quad 4\dfrac{1}{6} - 3\dfrac{1}{4}$

FRACTIONS

Basic Math

Evaluate the below fractions to the lowest possible terms.

(826) $1\frac{2}{3} - \frac{2}{3}$

(827) $4\frac{4}{5} - 2\frac{3}{4}$

(828) $3\frac{6}{7} - \frac{3}{4}$

(829) $4\frac{2}{7} - 4\frac{1}{6}$

(830) $7 - \frac{1}{3}$

(831) $3\frac{4}{5} - 2\frac{2}{3}$

(832) $3\frac{4}{5} - 1\frac{1}{2}$

(833) $1\frac{2}{3} - \frac{5}{8}$

(834) $4\frac{5}{7} - 2\frac{3}{8}$

(835) $2\frac{2}{3} - \frac{4}{5}$

 **FRACTIONS**

Evaluate the below fractions to the lowest possible terms.

(836) $2\dfrac{1}{2} - \dfrac{1}{2}$

(837) $2\dfrac{5}{8} - 2\dfrac{5}{8}$

(838) $3\dfrac{1}{2} - \dfrac{5}{6}$

(839) $4\dfrac{3}{7} - 1\dfrac{3}{8}$

(840) $4\dfrac{3}{4} - 3\dfrac{5}{8}$

(841) $4\dfrac{1}{7} - 3\dfrac{1}{2} - \dfrac{1}{6} - \dfrac{2}{7}$

(842) $4\dfrac{4}{5} - 1\dfrac{1}{6} - \dfrac{1}{2} - \dfrac{2}{7}$

(843) $3\dfrac{7}{8} - \dfrac{1}{4} - 1\dfrac{1}{2} - 1\dfrac{1}{5}$

(844) $2\dfrac{3}{8} - \dfrac{1}{8} - 1 - \dfrac{2}{3}$

(845) $4\dfrac{2}{3} - \dfrac{7}{8} - \dfrac{5}{6} - 1\dfrac{1}{2}$

FRACTIONS

Evaluate the below fractions to the lowest possible terms.

(846) $1\dfrac{1}{3} - 0 - \dfrac{5}{8} - \dfrac{4}{7}$

(847) $3\dfrac{2}{7} - \dfrac{1}{2} - 1\dfrac{1}{6} - \dfrac{5}{6}$

(848) $7\dfrac{1}{2} - \dfrac{1}{4} - 1\dfrac{1}{8} - 1\dfrac{4}{5}$

(849) $6\dfrac{2}{3} - 2\dfrac{3}{4} - 2\dfrac{2}{5} - \dfrac{1}{5}$

(850) $6 - 1\dfrac{1}{2} - \dfrac{3}{4} - 2\dfrac{5}{6}$

(851) $4\dfrac{1}{6} - 1\dfrac{1}{3} - 1\dfrac{1}{5} - \dfrac{1}{2}$

(852) $4\dfrac{1}{2} - 1\dfrac{2}{7} - \dfrac{2}{5} - \dfrac{1}{2}$

(853) $4\dfrac{6}{7} - 1\dfrac{5}{6} - \dfrac{3}{4} - 1\dfrac{3}{5}$

(854) $7\dfrac{1}{6} - \dfrac{4}{7} - 3\dfrac{1}{3} - 2\dfrac{3}{4}$

(855) $3\dfrac{5}{8} - \dfrac{2}{3} - 1\dfrac{2}{5} - \dfrac{7}{8}$

FRACTIONS

Basic Math

Evaluate the below fractions to the lowest possible terms.

(856) $4\dfrac{3}{4} - \dfrac{1}{3} - \dfrac{1}{4} - 3\dfrac{1}{2}$

(857) $2\dfrac{1}{8} - \dfrac{1}{6} - 1\dfrac{3}{4} - \dfrac{1}{7}$

(858) $2\dfrac{1}{6} - \dfrac{4}{7} - \dfrac{1}{2} - \dfrac{1}{2}$

(859) $3\dfrac{1}{3} - \dfrac{2}{7} - \dfrac{1}{2} - 1\dfrac{5}{8}$

(860) $8\dfrac{2}{3} - \dfrac{1}{4} - 1\dfrac{1}{3} - 3\dfrac{1}{3}$

(861) $2 - \dfrac{3}{8} - 0 - \dfrac{4}{5}$

(862) $4\dfrac{5}{7} - 1\dfrac{5}{6} - \dfrac{1}{2} - \dfrac{5}{7}$

(863) $4\dfrac{2}{3} - \dfrac{1}{4} - 1\dfrac{2}{7} - \dfrac{4}{5}$

(864) $4\dfrac{7}{8} - 1\dfrac{2}{3} - 2 - \dfrac{5}{7}$

(865) $8 - \dfrac{1}{6} - 4\dfrac{1}{2} - 1\dfrac{3}{4}$

FRACTIONS

Basic Math

Evaluate the below fractions to the lowest possible terms.

(866) $6\dfrac{1}{2} + 1\dfrac{5}{9} - 1\dfrac{1}{4} - 6\dfrac{7}{10}$

(867) $5\dfrac{7}{8} + 4\dfrac{5}{6} + 1\dfrac{3}{8} - 6\dfrac{3}{11}$

(868) $5\dfrac{2}{3} - 2\dfrac{2}{5} + 5\dfrac{7}{10} + 1\dfrac{1}{2}$

(869) $4\dfrac{1}{7} + 11\dfrac{6}{7} - 2\dfrac{10}{11} - 3\dfrac{5}{6}$

(870) $6\dfrac{2}{5} + 2\dfrac{7}{12} + \dfrac{9}{11} - 2\dfrac{1}{6}$

(871) $3\dfrac{1}{3} + 2\dfrac{6}{7} - 4\dfrac{3}{10} + \dfrac{3}{4}$

(872) $11\dfrac{1}{6} - 5\dfrac{3}{11} + \dfrac{1}{4} - 2\dfrac{1}{7}$

(873) $1\dfrac{3}{8} + 4\dfrac{1}{2} + 2\dfrac{1}{2} - 1\dfrac{4}{9}$

(874) $2\dfrac{1}{11} + 1\dfrac{4}{5} - 1\dfrac{3}{8} + 1\dfrac{2}{5}$

(875) $\dfrac{4}{5} + 1\dfrac{1}{3} + 6\dfrac{1}{2} - \dfrac{1}{4}$

FRACTIONS

Basic Math

Evaluate the below fractions to the lowest possible terms.

(876) $\quad 1\dfrac{9}{10} + \dfrac{3}{4} - 1\dfrac{3}{4} - 0$

(877) $\quad 6\dfrac{8}{11} - 1\dfrac{7}{12} - 2 + 5\dfrac{3}{11}$

(878) $\quad 2\dfrac{2}{5} + \dfrac{1}{12} + 4\dfrac{1}{4} - 5\dfrac{4}{9}$

(879) $\quad \dfrac{1}{11} + 2\dfrac{5}{6} + 5\dfrac{11}{12} + \dfrac{4}{5}$

(880) $\quad 5\dfrac{1}{12} + 1\dfrac{1}{3} + 6\dfrac{5}{11} + 3\dfrac{1}{4}$

(881) $\quad 4\dfrac{1}{8} + 2\dfrac{5}{8} + 1\dfrac{1}{2} + 3\dfrac{5}{8}$

(882) $\quad 1\dfrac{5}{6} + 6\dfrac{1}{3} - 1\dfrac{3}{8} + 5\dfrac{11}{12}$

(883) $\quad 3\dfrac{5}{6} + 5\dfrac{1}{5} + \dfrac{1}{3} - 2\dfrac{7}{10}$

(884) $\quad 5\dfrac{6}{11} + 6\dfrac{2}{5} - \dfrac{3}{10} + 5\dfrac{1}{10}$

(885) $\quad 4 - 2\dfrac{5}{6} + 5\dfrac{9}{10} + 5\dfrac{7}{9}$

FRACTIONS

Basic Math

Evaluate the below fractions to the lowest possible terms.

(886) $4\frac{5}{8} - 2\frac{7}{9} + 1\frac{4}{9} - 2\frac{1}{4}$

(887) $3\frac{9}{10} - \frac{4}{5} - 2\frac{1}{4} - \frac{1}{10}$

(888) $6\frac{6}{7} + 2\frac{3}{11} + 3\frac{2}{11} - 6\frac{4}{9}$

(889) $2\frac{3}{8} + 3\frac{7}{8} - 5\frac{1}{9} + \frac{1}{12}$

(890) $3\frac{1}{3} + 6\frac{1}{5} + 6\frac{7}{10} + 4\frac{1}{8}$

(891) $4\frac{2}{5} + 1\frac{1}{5} + 6\frac{1}{3} + 2\frac{1}{10}$

(892) $4\frac{5}{8} - 1\frac{7}{8} - 1\frac{9}{11} + 11\frac{1}{2}$

(893) $2 + 4\frac{3}{7} + 4\frac{7}{10} + 5\frac{3}{11}$

(894) $10 + 4\frac{1}{12} - \frac{8}{11} + 1\frac{5}{7}$

(895) $1\frac{11}{12} - \frac{5}{7} + 5\frac{2}{7} - \frac{1}{6}$

 **FRACTIONS**

Basic Math

Find the product of each of the below fractions given.

(896) $1\dfrac{3}{7} \times \dfrac{1}{3}$

(897) $2\dfrac{2}{3} \times 1\dfrac{5}{6}$

(898) $\dfrac{7}{6} \times \dfrac{1}{3}$

(899) $1\dfrac{1}{2} \times \dfrac{1}{4}$

(900) $9 \times \dfrac{5}{7}$

(901) $1\dfrac{2}{5} \times \dfrac{5}{3}$

(902) $3 \times \dfrac{3}{4}$

(903) $4\dfrac{3}{10} \times \dfrac{3}{4}$

(904) $2\dfrac{1}{10} \times \dfrac{1}{2}$

(905) $\dfrac{1}{4} \times \dfrac{1}{6}$

 **FRACTIONS**

Basic Math

Find the product of each of the below fractions given.

(906) $1\dfrac{1}{2} \times \dfrac{3}{5}$

(907) $3\dfrac{8}{9} \times \dfrac{1}{2}$

(908) $\dfrac{3}{8} \times \dfrac{5}{3}$

(909) $4\dfrac{1}{3} \times \dfrac{6}{5}$

(910) $1\dfrac{3}{5} \times 2$

(911) $2\dfrac{1}{2} \times \dfrac{15}{8}$

(912) $\dfrac{1}{4} \times \dfrac{1}{9}$

(913) $3\dfrac{5}{8} \times 3\dfrac{3}{4}$

(914) $\dfrac{3}{4} \times \dfrac{4}{7}$

(915) $\dfrac{4}{3} \times \dfrac{17}{9}$

 **FRACTIONS**

Basic Math

Find the product of each of the below fractions given.

(916) $2\dfrac{1}{4} \times \dfrac{8}{5}$

(917) $1\dfrac{5}{6} \times 4\dfrac{1}{6}$

(918) $2 \times \dfrac{1}{5}$

(919) $2\dfrac{7}{10} \times \dfrac{3}{10}$

(920) $\dfrac{1}{2} \times \dfrac{1}{2}$

(921) $\dfrac{19}{10} \times \dfrac{3}{4} \times \dfrac{2}{5}$

(922) $1\dfrac{4}{9} \times 5\dfrac{5}{6} \times \dfrac{4}{5}$

(923) $\dfrac{1}{7} \times \dfrac{5}{9} \times \dfrac{3}{5}$

(924) $5\dfrac{4}{9} \times 4\dfrac{3}{4} \times 4\dfrac{3}{4}$

(925) $2\dfrac{1}{2} \times 5\dfrac{5}{6} \times 5\dfrac{1}{2}$

FRACTIONS

Basic Math

Find the product of each of the below fractions given.

(926) $2 \times \dfrac{2}{5} \times \dfrac{2}{3}$

(927) $5\dfrac{1}{3} \times 8 \times \dfrac{7}{8}$

(928) $3\dfrac{1}{2} \times 5\dfrac{1}{10} \times \dfrac{1}{4}$

(929) $4\dfrac{3}{10} \times \dfrac{5}{4} \times \dfrac{3}{2}$

(930) $1\dfrac{1}{9} \times 2 \times \dfrac{4}{5}$

(931) $2\dfrac{4}{5} \times \dfrac{8}{5} \times \dfrac{17}{9}$

(932) $2 \times 3\dfrac{5}{6} \times \dfrac{3}{2}$

(933) $2 \times \dfrac{3}{2} \times \dfrac{3}{2}$

(934) $1\dfrac{1}{10} \times 5\dfrac{3}{10} \times \dfrac{13}{9}$

(935) $4\dfrac{3}{7} \times 5\dfrac{1}{8} \times \dfrac{1}{7}$

 **FRACTIONS**   **Basic Math**

Find the product of each of the below fractions given.

(936) $2\dfrac{1}{6} \times \dfrac{1}{6} \times \dfrac{1}{5}$ (937) $5\dfrac{1}{2} \times \dfrac{3}{8} \times \dfrac{12}{7}$

(938) $4\dfrac{1}{2} \times 4\dfrac{1}{7} \times 2\dfrac{1}{2}$ (939) $1\dfrac{7}{10} \times 3\dfrac{7}{8} \times 3\dfrac{2}{3}$

(940) $5\dfrac{1}{6} \times 3\dfrac{5}{6} \times \dfrac{5}{3}$ (941) $2\dfrac{9}{10} \times \dfrac{5}{6} \times \dfrac{1}{2}$

(942) $4\dfrac{1}{2} \times 1\dfrac{7}{8} \times \dfrac{5}{6}$ (943) $5\dfrac{7}{9} \times 2 \times \dfrac{4}{3}$

(944) $\dfrac{1}{3} \times \dfrac{13}{9} \times \dfrac{12}{7}$ (945) $1\dfrac{1}{2} \times 2 \times \dfrac{4}{3}$

 **FRACTIONS**

Basic Math

Find the product of each of the below fractions given.

(946) $6\frac{8}{9} \times 3\frac{4}{11} \times -3\frac{4}{5} \times \frac{1}{4}$

(947) $1\frac{3}{4} \times \frac{5}{6} \times \frac{3}{5} \times -\frac{13}{9}$

(948) $1\frac{1}{2} \times 1\frac{1}{5} \times -\frac{4}{3} \times -\frac{1}{3}$

(949) $12 \times -5 \times -2 \times -\frac{1}{3}$

(950) $-1\frac{5}{12} \times 9 \times \frac{3}{8} \times \frac{5}{6}$

(951) $4\frac{5}{7} \times 2\frac{4}{11} \times -3\frac{2}{11} \times -\frac{11}{6}$

(952) $4\frac{4}{7} \times 2\frac{3}{5} \times -9\frac{1}{6} \times \frac{7}{8}$

(953) $3\frac{2}{7} \times 2\frac{1}{2} \times -\frac{1}{2} \times \frac{17}{10}$

(954) $10\frac{7}{9} \times 6\frac{1}{10} \times 6\frac{9}{11} \times -\frac{4}{3}$

(955) $3\frac{1}{10} \times -11 \times 2\frac{3}{8} \times 6\frac{1}{11}$

Find the product of each of the below fractions given.

(956) $\quad -2 \times 6 \times -1 \times \dfrac{3}{5}$

(957) $\quad -3\dfrac{11}{12} \times \dfrac{5}{11} \times \dfrac{5}{9} \times -\dfrac{5}{4}$

(958) $\quad 1\dfrac{1}{8} \times -\dfrac{3}{5} \times -\dfrac{7}{4} \times \dfrac{1}{2}$

(959) $\quad 2\dfrac{1}{8} \times -3\dfrac{7}{11} \times 5\dfrac{6}{7} \times 2$

(960) $\quad 3\dfrac{3}{10} \times 1\dfrac{5}{7} \times -\dfrac{7}{11} \times -\dfrac{11}{6}$

(961) $\quad 4\dfrac{3}{4} \times 2\dfrac{1}{6} \times -\dfrac{5}{6} \times -\dfrac{11}{6}$

(962) $\quad 11 \times 5\dfrac{7}{12} \times -\dfrac{5}{3} \times -\dfrac{2}{5}$

(963) $\quad -2\dfrac{1}{5} \times 6\dfrac{1}{10} \times 3\dfrac{7}{9} \times \dfrac{1}{2}$

(964) $\quad -1\dfrac{7}{12} \times 3\dfrac{3}{5} \times \dfrac{1}{2} \times -\dfrac{5}{8}$

(965) $\quad -1\dfrac{1}{3} \times 5\dfrac{1}{4} \times -\dfrac{2}{3} \times \dfrac{5}{3}$

FRACTIONS

Find the product of each of the below fractions given.

(966) $-1\dfrac{4}{5} \times -2 \times -1 \times \dfrac{1}{3}$

(967) $-7 \times \dfrac{6}{5} \times -\dfrac{1}{4} \times \dfrac{1}{3}$

(968) $-\dfrac{1}{2} \times -\dfrac{3}{2} \times \dfrac{2}{11} \times \dfrac{10}{7}$

(969) $5\dfrac{1}{2} \times -\dfrac{3}{4} \times -\dfrac{1}{2} \times -\dfrac{7}{5}$

(970) $-2 \times 5\dfrac{11}{12} \times 2\dfrac{11}{12} \times -\dfrac{6}{5}$

(971) $4\dfrac{4}{9} \times 5\dfrac{8}{9}$

(972) $3\dfrac{4}{9} \times \dfrac{1}{3}$

(973) $3\dfrac{5}{6} \times 3\dfrac{5}{6}$

(974) $2\dfrac{2}{5} \times 3\dfrac{2}{3}$

(975) $1\dfrac{5}{6} \times 2\dfrac{7}{8}$

FRACTIONS

Basic Math

Find the product of each of the below fractions given.

(976) $4\dfrac{1}{8} \times 2\dfrac{1}{2}$

(977) $2\dfrac{9}{10} \times 4\dfrac{6}{7}$

(978) $5\dfrac{1}{6} \times 4\dfrac{2}{7}$

(979) $3\dfrac{3}{10} \times 6\dfrac{6}{7}$

(980) $5\dfrac{1}{4} \times 1\dfrac{7}{8}$

(981) $2\dfrac{6}{7} \times 4\dfrac{3}{10}$

(982) $5\dfrac{7}{9} \times 1\dfrac{1}{10}$

(983) $3\dfrac{1}{2} \times \dfrac{1}{2}$

(984) $4\dfrac{3}{4} \times \dfrac{3}{8}$

(985) $5\dfrac{1}{4} \times 4\dfrac{4}{7}$

FRACTIONS

Basic Math

Find the product of each of the below fractions given.

(986) $1\dfrac{6}{7} \times 4\dfrac{1}{8}$

(987) $4\dfrac{9}{10} \times 5\dfrac{4}{5}$

(988) $5\dfrac{1}{9} \times \dfrac{3}{4}$

(989) $5\dfrac{5}{6} \times 4\dfrac{1}{4}$

(990) $4\dfrac{5}{7} \times \dfrac{1}{2}$

(991) $5\dfrac{1}{2} \times 4\dfrac{3}{7}$

(992) $3\dfrac{1}{4} \times 4\dfrac{1}{6}$

(993) $5\dfrac{3}{4} \times 1\dfrac{3}{4}$

(994) $5\dfrac{4}{9} \times \dfrac{1}{7}$

(995) $1\dfrac{3}{10} \times 4\dfrac{5}{6}$

Find the product of each of the below fractions given.

(996) $6 \times 5\dfrac{1}{9} \times 2\dfrac{1}{2}$

(997) $4\dfrac{4}{11} \times 1\dfrac{9}{11} \times \dfrac{3}{4}$

(998) $5\dfrac{3}{11} \times 6\dfrac{1}{4} \times \dfrac{5}{6}$

(999) $3\dfrac{7}{12} \times 5\dfrac{3}{10} \times 2\dfrac{4}{7}$

(1000) $3\dfrac{9}{10} \times 3\dfrac{6}{11} \times \dfrac{4}{11}$

(1001) $3\dfrac{5}{9} \times 1\dfrac{1}{3} \times 3\dfrac{3}{11}$

(1002) $1\dfrac{11}{12} \times 6\dfrac{1}{10} \times 2\dfrac{1}{6}$

(1003) $4\dfrac{1}{2} \times 4\dfrac{1}{3} \times \dfrac{1}{2}$

(1004) $1\dfrac{3}{10} \times 4\dfrac{1}{4} \times \dfrac{1}{6}$

(1005) $4\dfrac{5}{8} \times 4\dfrac{3}{8} \times 1\dfrac{3}{8}$

 **FRACTIONS**

Find the product of each of the below fractions given.

(1006) $6\dfrac{5}{6} \times \dfrac{7}{12} \times \dfrac{7}{10}$

(1007) $3\dfrac{4}{7} \times 4\dfrac{4}{5} \times 6\dfrac{1}{6}$

(1008) $6\dfrac{4}{7} \times 1\dfrac{7}{8} \times 5\dfrac{1}{2}$

(1009) $1\dfrac{3}{10} \times 6\dfrac{1}{4} \times \dfrac{8}{11}$

(1010) $4\dfrac{3}{7} \times 6\dfrac{5}{6} \times 9$

(1011) $2\dfrac{1}{7} \times 3\dfrac{1}{6} \times \dfrac{5}{8}$

(1012) $1\dfrac{1}{2} \times 4\dfrac{11}{12} \times \dfrac{2}{5}$

(1013) $2\dfrac{3}{5} \times 5\dfrac{1}{6} \times 4\dfrac{3}{8}$

(1014) $3\dfrac{4}{9} \times 4\dfrac{5}{11} \times \dfrac{7}{12}$

(1015) $2\dfrac{7}{8} \times 3\dfrac{9}{11} \times 2\dfrac{5}{9}$

FRACTIONS

Basic Math

Find the product of each of the below fractions given.

(1016) $5\frac{5}{7} \times 2\frac{5}{8} \times 5\frac{3}{4}$

(1017) $6\frac{5}{7} \times 2\frac{1}{12} \times 6\frac{5}{12}$

(1018) $1\frac{7}{10} \times 5\frac{1}{6} \times 3\frac{10}{11}$

(1019) $9\frac{2}{3} \times 6\frac{8}{11} \times 3\frac{1}{2}$

(1020) $3\frac{1}{2} \times 1\frac{1}{8} \times \frac{5}{11}$

(1021) $\frac{1}{2} \times \frac{1}{3}$

(1022) $\frac{10}{7} \times \frac{2}{3}$

(1023) $\frac{9}{5} \times \frac{9}{5}$

(1024) $\frac{6}{5} \times \frac{15}{8}$

(1025) $2 \times \frac{1}{9}$

 **FRACTIONS**

Basic Math

Find the product of each of the below fractions given.

(1026) $\dfrac{2}{3} \times \dfrac{1}{2}$

(1027) $6 \times \dfrac{7}{9}$

(1028) $2 \times \dfrac{3}{2}$

(1029) $0 \times \dfrac{2}{9}$

(1030) $\dfrac{13}{8} \times \dfrac{2}{3}$

(1031) $\dfrac{1}{3} \times \dfrac{3}{2}$

(1032) $8 \times \dfrac{4}{3}$

(1033) $\dfrac{5}{4} \times \dfrac{17}{9}$

(1034) $\dfrac{15}{8} \times \dfrac{13}{7}$

(1035) $\dfrac{3}{2} \times \dfrac{3}{4}$

FRACTIONS

Find the product of each of the below fractions given.

(1036) $2 \times \dfrac{7}{8}$

(1037) $2 \times \dfrac{13}{8}$

(1038) $\dfrac{3}{2} \times \dfrac{1}{2}$

(1039) $\dfrac{7}{9} \times \dfrac{8}{5}$

(1040) $\dfrac{5}{3} \times \dfrac{2}{7}$

(1041) $2 \times \dfrac{2}{3}$

(1042) $2 \times \dfrac{1}{3}$

(1043) $\dfrac{9}{8} \times \dfrac{3}{2}$

(1044) $3 \times \dfrac{19}{10}$

(1045) $\dfrac{13}{9} \times \dfrac{12}{7}$

FRACTIONS

Basic Math

Find the product of each of the below fractions given.

(1046) $\dfrac{1}{2} \times \dfrac{5}{7} \times \dfrac{13}{7}$

(1047) $2 \times \dfrac{3}{4} \times \dfrac{10}{9}$

(1048) $2 \times \dfrac{4}{3} \times \dfrac{2}{7}$

(1049) $2 \times \dfrac{3}{2} \times \dfrac{19}{10}$

(1050) $2 \times \dfrac{1}{3} \times \dfrac{1}{2}$

(1051) $2 \times \dfrac{8}{7} \times \dfrac{6}{5}$

(1052) $\dfrac{1}{6} \times \dfrac{11}{10} \times \dfrac{4}{3}$

(1053) $\dfrac{9}{5} \times \dfrac{17}{10} \times \dfrac{3}{2}$

(1054) $2 \times \dfrac{3}{4} \times \dfrac{3}{2}$

(1055) $2 \times \dfrac{5}{7} \times \dfrac{13}{9}$

FRACTIONS

Basic Math

Find the product of each of the below fractions given.

(1056) $\dfrac{5}{3} \times \dfrac{5}{6} \times \dfrac{6}{5}$

(1057) $\dfrac{9}{5} \times \dfrac{16}{9} \times \dfrac{9}{5}$

(1058) $\dfrac{8}{7} \times \dfrac{8}{5} \times \dfrac{5}{3}$

(1059) $2 \times \dfrac{1}{2} \times \dfrac{2}{3}$

(1060) $0 \times \dfrac{14}{9} \times \dfrac{6}{5}$

(1061) $\dfrac{5}{6} \times \dfrac{3}{2} \times \dfrac{3}{2}$

(1062) $\dfrac{1}{2} \times \dfrac{5}{7} \times \dfrac{2}{3}$

(1063) $\dfrac{10}{9} \times \dfrac{1}{4} \times \dfrac{9}{8}$

(1064) $5 \times \dfrac{11}{9} \times \dfrac{6}{7}$

(1065) $2 \times 7 \times \dfrac{11}{6}$

Find the product of each of the below fractions given.

(1066) $\dfrac{3}{2} \times \dfrac{4}{3} \times \dfrac{9}{5}$

(1067) $2 \times \dfrac{1}{2} \times \dfrac{10}{9}$

(1068) $10 \times \dfrac{2}{7} \times \dfrac{2}{7}$

(1069) $\dfrac{5}{6} \times \dfrac{3}{5} \times \dfrac{6}{7}$

(1070) $\dfrac{11}{7} \times \dfrac{3}{2} \times \dfrac{9}{10}$

FRACTIONS

Basic Math

Find the quotient of each of the below fractions given.

(1071) $\dfrac{1}{10} \div \dfrac{5}{3}$

(1072) $5\dfrac{4}{7} \div 5\dfrac{3}{4}$

(1073) $10 \div \dfrac{4}{5}$

(1074) $2 \div \dfrac{1}{6}$

(1075) $\dfrac{1}{4} \div \dfrac{3}{2}$

(1076) $\dfrac{7}{4} \div \dfrac{2}{5}$

(1077) $4\dfrac{4}{7} \div \dfrac{3}{2}$

(1078) $5\dfrac{1}{2} \div 3\dfrac{1}{6}$

(1079) $\dfrac{7}{5} \div 3\dfrac{2}{5}$

(1080) $\dfrac{6}{5} \div \dfrac{3}{7}$

 **FRACTIONS**

Find the quotient of each of the below fractions given.

(1081) $\dfrac{1}{3} \div \dfrac{1}{4}$ (1082) $\dfrac{5}{7} \div \dfrac{2}{3}$

(1083) $\dfrac{5}{3} \div 1\dfrac{1}{2}$ (1084) $\dfrac{2}{3} \div \dfrac{4}{7}$

(1085) $\dfrac{3}{5} \div \dfrac{1}{3}$ (1086) $5\dfrac{1}{2} \div \dfrac{1}{6}$

(1087) $4\dfrac{1}{2} \div 2$ (1088) $2 \div 2\dfrac{1}{6}$

(1089) $2\dfrac{7}{8} \div \dfrac{1}{2}$ (1090) $\dfrac{7}{4} \div 2$

FRACTIONS

Basic Math

Find the quotient of each of the below fractions given.

(1091) $1 \div \dfrac{4}{3}$

(1092) $\dfrac{2}{5} \div 4\dfrac{2}{5}$

(1093) $2\dfrac{3}{5} \div 2$

(1094) $2 \div 2\dfrac{4}{7}$

(1095) $9 \div \dfrac{10}{9}$

(1096) $\dfrac{2}{3} \div \dfrac{3}{4}$

(1097) $2\dfrac{2}{7} \div \dfrac{1}{4}$

(1098) $4\dfrac{5}{6} \div \dfrac{3}{10}$

(1099) $4\dfrac{1}{5} \div 5$

(1100) $\dfrac{3}{10} \div \dfrac{1}{8}$

FRACTIONS

Find the quotient of each of the below fractions given.

(1101) $2\dfrac{3}{7} \div 5\dfrac{3}{8}$

(1102) $4\dfrac{1}{2} \div 4\dfrac{1}{3}$

(1103) $3\dfrac{2}{7} \div 4\dfrac{1}{5}$

(1104) $5\dfrac{7}{9} \div 4\dfrac{7}{9}$

(1105) $3\dfrac{1}{8} \div 2\dfrac{5}{8}$

(1106) $2\dfrac{4}{5} \div 1\dfrac{1}{2}$

(1107) $3\dfrac{3}{10} \div 4\dfrac{1}{3}$

(1108) $5\dfrac{3}{4} \div 5\dfrac{2}{3}$

(1109) $1\dfrac{1}{2} \div 5\dfrac{1}{3}$

(1110) $5\dfrac{1}{6} \div 5\dfrac{1}{6}$

 **FRACTIONS**

Find the quotient of each of the below fractions given.

(1111) $1\dfrac{4}{7} \div \dfrac{5}{9}$

(1112) $2\dfrac{1}{2} \div 4\dfrac{1}{2}$

(1113) $4\dfrac{5}{9} \div 5\dfrac{1}{2}$

(1114) $4\dfrac{1}{2} \div 3\dfrac{7}{9}$

(1115) $9\dfrac{1}{2} \div 1\dfrac{3}{5}$

(1116) $5\dfrac{1}{2} \div 3\dfrac{1}{9}$

(1117) $\dfrac{3}{4} \div 2\dfrac{1}{5}$

(1118) $3\dfrac{3}{7} \div 3\dfrac{1}{7}$

(1119) $4\dfrac{1}{4} \div 1\dfrac{1}{10}$

(1120) $2\dfrac{2}{3} \div 3\dfrac{2}{5}$

FRACTIONS

Find the quotient of each of the below fractions given.

(1121) $\dfrac{2}{3} \div 2$

(1122) $2 \div \dfrac{1}{2}$

(1123) $\dfrac{3}{2} \div \dfrac{11}{7}$

(1124) $\dfrac{10}{9} \div \dfrac{4}{5}$

(1125) $\dfrac{2}{3} \div \dfrac{1}{2}$

(1126) $\dfrac{3}{2} \div \dfrac{1}{4}$

(1127) $\dfrac{1}{2} \div \dfrac{10}{7}$

(1128) $\dfrac{3}{4} \div \dfrac{1}{4}$

(1129) $3 \div \dfrac{4}{3}$

(1130) $\dfrac{1}{3} \div \dfrac{1}{2}$

 **FRACTIONS**

Find the quotient of each of the below fractions given.

(1131) $\dfrac{9}{5} \div \dfrac{6}{5}$ (1132) $\dfrac{11}{8} \div \dfrac{8}{9}$

(1133) $\dfrac{19}{10} \div 2$ (1134) $\dfrac{3}{2} \div \dfrac{3}{2}$

(1135) $\dfrac{5}{3} \div \dfrac{5}{3}$ (1136) $\dfrac{5}{9} \div \dfrac{4}{3}$

(1137) $\dfrac{9}{7} \div \dfrac{14}{9}$ (1138) $\dfrac{1}{5} \div \dfrac{9}{8}$

(1139) $\dfrac{2}{3} \div \dfrac{2}{3}$ (1140) $\dfrac{5}{6} \div \dfrac{5}{4}$

FRACTIONS

Find the quotient of each of the below fractions given.

(1141) $\dfrac{5}{4} \div \dfrac{5}{6}$

(1142) $\dfrac{1}{6} \div \dfrac{5}{4}$

(1143) $\dfrac{4}{5} \div \dfrac{11}{8}$

(1144) $2 \div \dfrac{2}{9}$

(1145) $\dfrac{1}{8} \div \dfrac{1}{2}$

FRACTIONS

Basic Math Answer Keys

Answer Key

(1) $\dfrac{2}{3}$ (2) $\dfrac{1}{3}$ (3) $\dfrac{3}{7}$ (4) $\dfrac{1}{4}$

(5) $\dfrac{3}{8}$ (6) $\dfrac{2}{3}$ (7) $\dfrac{1}{8}$ (8) $\dfrac{3}{7}$

(9) $\dfrac{7}{8}$ (10) $\dfrac{5}{8}$ (11) $\dfrac{2}{3}$ (12) $\dfrac{6}{7}$

(13) $\dfrac{2}{5}$ (14) $\dfrac{1}{6}$ (15) $\dfrac{3}{4}$ (16) $\dfrac{1}{4}$

(17) $\dfrac{1}{2}$ (18) $\dfrac{1}{3}$ (19) $\dfrac{4}{5}$ (20) $\dfrac{2}{7}$

(21) $\dfrac{1}{3}$ (22) $\dfrac{1}{4}$ (23) $\dfrac{1}{2}$ (24) $\dfrac{3}{8}$

(25) $\dfrac{1}{2}$ (26) $\dfrac{1}{4}$ (27) $\dfrac{3}{5}$ (28) $\dfrac{4}{5}$

(29) $\dfrac{3}{5}$ (30) $\dfrac{1}{2}$ (31) $\dfrac{1}{9}$ (32) $\dfrac{7}{8}$

(33) $\dfrac{5}{8}$ (34) $\dfrac{1}{4}$ (35) $\dfrac{1}{7}$ (36) $\dfrac{3}{10}$

FRACTIONS

Basic Math Answer Keys

(37) $\dfrac{1}{4}$ (38) $\dfrac{2}{3}$ (39) $\dfrac{7}{10}$ (40) $\dfrac{4}{7}$

(41) $\dfrac{1}{3}$ (42) $\dfrac{2}{7}$ (43) $\dfrac{4}{9}$ (44) $\dfrac{3}{7}$

(45) $\dfrac{1}{3}$ (46) $\dfrac{1}{2}$ (47) $\dfrac{5}{8}$ (48) $\dfrac{4}{5}$

(49) $\dfrac{1}{2}$ (50) $\dfrac{1}{3}$ (51) $\dfrac{5}{13}$ (52) $\dfrac{1}{2}$

(53) $\dfrac{4}{15}$ (54) $\dfrac{2}{3}$ (55) $\dfrac{1}{15}$ (56) $\dfrac{2}{5}$

(57) $\dfrac{2}{13}$ (58) $\dfrac{11}{13}$ (59) $\dfrac{13}{15}$ (60) $\dfrac{11}{15}$

(61) $\dfrac{5}{6}$ (62) $\dfrac{7}{15}$ (63) $\dfrac{7}{8}$ (64) $\dfrac{5}{6}$

(65) $\dfrac{5}{8}$ (66) $\dfrac{13}{14}$ (67) $\dfrac{2}{7}$ (68) $\dfrac{2}{15}$

(69) $\dfrac{13}{15}$ (70) $\dfrac{3}{5}$ (71) $\dfrac{10}{13}$ (72) $\dfrac{6}{13}$

FRACTIONS

Basic Math Answer Keys

(73) $\dfrac{7}{9}$ (74) $\dfrac{1}{2}$ (75) $\dfrac{2}{3}$ (76) $\dfrac{2}{7}$

(77) $\dfrac{5}{12}$ (78) $\dfrac{4}{5}$ (79) $\dfrac{7}{13}$ (80) $\dfrac{2}{7}$

(81) $\dfrac{1}{4}$ (82) $\dfrac{5}{7}$ (83) $\dfrac{4}{9}$ (84) $\dfrac{9}{11}$

(85) $\dfrac{5}{13}$ (86) $\dfrac{7}{12}$ (87) $\dfrac{1}{15}$ (88) $\dfrac{1}{12}$

(89) $\dfrac{1}{13}$ (90) $\dfrac{4}{5}$ (91) $\dfrac{1}{3}$ (92) $\dfrac{2}{13}$

(93) $\dfrac{4}{11}$ (94) $\dfrac{11}{12}$ (95) $\dfrac{8}{15}$ (96) $\dfrac{1}{9}$

(97) $\dfrac{9}{10}$ (98) $\dfrac{1}{3}$ (99) $\dfrac{11}{14}$ (100) $\dfrac{10}{11}$

(101) $8\dfrac{1}{2}$ (102) $3\dfrac{1}{2}$ (103) $7\dfrac{1}{3}$ (104) $2\dfrac{2}{3}$

(105) $7\dfrac{3}{5}$ (106) $1\dfrac{1}{2}$ (107) $6\dfrac{2}{3}$ (108) $4\dfrac{3}{5}$

FRACTIONS

(109) $5\dfrac{1}{5}$ (110) $1\dfrac{1}{6}$ (111) $5\dfrac{3}{4}$ (112) $8\dfrac{1}{6}$

(113) $4\dfrac{1}{4}$ (114) $4\dfrac{1}{5}$ (115) $4\dfrac{1}{7}$ (116) $4\dfrac{1}{2}$

(117) $7\dfrac{2}{3}$ (118) $2\dfrac{1}{4}$ (119) $1\dfrac{2}{5}$ (120) $4\dfrac{1}{7}$

(121) $1\dfrac{1}{7}$ (122) $7\dfrac{3}{4}$ (123) $4\dfrac{2}{5}$ (124) $8\dfrac{1}{2}$

(125) $2\dfrac{1}{3}$ (126) $10\dfrac{1}{8}$ (127) $5\dfrac{1}{9}$ (128) $1\dfrac{3}{4}$

(129) $9\dfrac{1}{2}$ (130) $1\dfrac{1}{2}$ (131) $12\dfrac{3}{4}$ (132) $10\dfrac{4}{7}$

(133) $12\dfrac{1}{3}$ (134) $2\dfrac{3}{5}$ (135) $11\dfrac{1}{2}$ (136) $8\dfrac{1}{6}$

(137) $5\dfrac{1}{7}$ (138) $10\dfrac{3}{5}$ (139) $11\dfrac{1}{5}$ (140) $1\dfrac{3}{4}$

(141) $2\dfrac{1}{4}$ (142) $1\dfrac{1}{8}$ (143) $7\dfrac{1}{5}$ (144) $3\dfrac{1}{4}$

 **FRACTIONS**

Basic Math Answer Keys

(145) $8\frac{3}{5}$ (146) $3\frac{1}{3}$ (147) $9\frac{1}{2}$ (148) $9\frac{1}{2}$

(149) $7\frac{3}{7}$ (150) $8\frac{5}{6}$ (151) $4\frac{2}{3}$ (152) $10\frac{1}{5}$

(153) $4\frac{1}{3}$ (154) $8\frac{1}{5}$ (155) $6\frac{1}{5}$ (156) $3\frac{1}{2}$

(157) $10\frac{2}{5}$ (158) $12\frac{2}{9}$ (159) $11\frac{1}{3}$ (160) $3\frac{4}{5}$

(161) $2\frac{1}{3}$ (162) $3\frac{1}{7}$ (163) $12\frac{1}{10}$ (164) $8\frac{1}{2}$

(165) $5\frac{5}{6}$ (166) $7\frac{1}{5}$ (167) $5\frac{2}{3}$ (168) $9\frac{2}{7}$

(169) $7\frac{1}{6}$ (170) $12\frac{2}{3}$ (171) $1\frac{1}{10}$ (172) $5\frac{1}{4}$

(173) $9\frac{1}{2}$ (174) $11\frac{1}{7}$ (175) $9\frac{2}{5}$ (176) $\frac{9}{5}$

(177) $\frac{9}{4}$ (178) $\frac{3}{2}$ (179) $\frac{7}{4}$ (180) $\frac{6}{5}$

FRACTIONS

Basic Math Answer Keys

(181) $\dfrac{10}{9}$ (182) $\dfrac{7}{4}$ (183) $\dfrac{5}{2}$ (184) $\dfrac{7}{5}$

(185) $\dfrac{5}{4}$ (186) $\dfrac{9}{5}$ (187) $\dfrac{3}{2}$ (188) $\dfrac{5}{3}$

(189) $\dfrac{7}{5}$ (190) $\dfrac{8}{5}$ (191) $\dfrac{9}{8}$ (192) $\dfrac{7}{3}$

(193) $\dfrac{7}{6}$ (194) $\dfrac{7}{3}$ (195) $\dfrac{5}{3}$ (196) $\dfrac{5}{2}$

(197) $\dfrac{4}{3}$ (198) $\dfrac{5}{4}$ (199) $\dfrac{6}{5}$ (200) $\dfrac{9}{8}$

(201) $\dfrac{19}{9}$ (202) $\dfrac{16}{3}$ (203) $\dfrac{8}{7}$ (204) $\dfrac{11}{7}$

(205) $\dfrac{13}{5}$ (206) $\dfrac{25}{18}$ (207) $\dfrac{7}{5}$ (208) $\dfrac{8}{5}$

(209) $\dfrac{3}{2}$ (210) $\dfrac{24}{19}$ (211) $\dfrac{6}{5}$ (212) $\dfrac{5}{2}$

(213) $\dfrac{17}{12}$ (214) $\dfrac{11}{4}$ (215) $\dfrac{17}{4}$ (216) $\dfrac{6}{5}$

FRACTIONS

Basic Math Answer Keys

(217) $\dfrac{4}{3}$ (218) $\dfrac{7}{4}$ (219) $\dfrac{22}{17}$ (220) $\dfrac{22}{19}$

(221) $\dfrac{21}{11}$ (222) $\dfrac{23}{18}$ (223) $\dfrac{17}{16}$ (224) $\dfrac{19}{8}$

(225) $\dfrac{23}{17}$ (226) $2\dfrac{2}{3}$ (227) $1\dfrac{1}{9}$ (228) $1\dfrac{1}{4}$

(229) $2\dfrac{1}{2}$ (230) $1\dfrac{1}{2}$ (231) $1\dfrac{1}{3}$ (232) $1\dfrac{1}{8}$

(233) $2\dfrac{1}{4}$ (234) $1\dfrac{1}{9}$ (235) $1\dfrac{3}{4}$ (236) $3\dfrac{1}{3}$

(237) $1\dfrac{1}{4}$ (238) $1\dfrac{1}{8}$ (239) $1\dfrac{1}{3}$ (240) $3\dfrac{1}{3}$

(241) $1\dfrac{1}{4}$ (242) $1\dfrac{1}{3}$ (243) $1\dfrac{1}{2}$ (244) $1\dfrac{3}{4}$

(245) $1\dfrac{1}{2}$ (246) $2\dfrac{1}{2}$ (247) $1\dfrac{1}{3}$ (248) $1\dfrac{2}{3}$

(249) $2\dfrac{1}{2}$ (250) $1\dfrac{2}{3}$ (251) $1\dfrac{2}{5}$ (252) $1\dfrac{1}{5}$

FRACTIONS

Basic Math Answer Keys

(253) $1\frac{4}{9}$ (254) $1\frac{1}{6}$ (255) $2\frac{2}{3}$ (256) $1\frac{1}{5}$

(257) $1\frac{3}{7}$ (258) $1\frac{1}{12}$ (259) $1\frac{2}{7}$ (260) $1\frac{3}{7}$

(261) $3\frac{1}{3}$ (262) $1\frac{1}{2}$ (263) $1\frac{3}{5}$ (264) $1\frac{2}{3}$

(265) $1\frac{1}{4}$ (266) $1\frac{1}{3}$ (267) $1\frac{1}{5}$ (268) $1\frac{1}{4}$

(269) $1\frac{2}{5}$ (270) $2\frac{3}{5}$ (271) $2\frac{2}{5}$ (272) $3\frac{1}{4}$

(273) $2\frac{1}{3}$ (274) $1\frac{1}{2}$ (275) $1\frac{6}{7}$ (276) 0.6

(277) 0.3   (278) 0.91   (279) $6.\overline{15}$ (280) 4.75

(281) 0.625   (282) 0.67   (283) 0.875   (284) $0.\overline{45}$

(285) $9.\overline{6}$ (286) 2.25   (287) 0.1   (288) 0.2

152

 **FRACTIONS**

Basic Math Answer Keys

(289) 0.5   (290) 3.5   (291) 0.$\overline{3}$ (292) 0.$\overline{15}$

(293) 0.$\overline{005}$ (294) 0.125   (295) 0.9   (296) 0.7

(297) 6.5   (298) 0.$\overline{6}$ (299) 6.752   (300) 8.88

(301) 0.88   (302) 0.01   (303) 0.33   (304) 1.15

(305) 0.31   (306) 5.5   (307) 8.12   (308) 1.6

(309) 1.3   (310) 0.25   (311) 0.42   (312) 0.01

(313) 0   (314) 3.5   (315) 5.12   (316) 0.12

(317) 0.37   (318) 0.1   (319) 9.2   (320) 0.7

(321) 0.62   (322) 0   (323) 4.41   (324) 9.71

 FRACTIONS

Basic Math
Answer Keys

(325) 0.69   (326) 0.42   (327) 0.9   (328) 0.009

(329) 6.75   (330) 0.524   (331) 0.625   (332) 7.75

(333) 0.01   (334) 4.5   (335) 6.725   (336) 0.008

(337) 0.001   (338) 1.535   (339) 0.004   (340) 0.544

(341) 0.754   (342) 6.74   (343) 0.167   (344) 0.005

(345) 0.5   (346) 0.25   (347) 0.96   (348) 0.875

(349) 0.2   (350) 0.7   (351) 2   (352) 1

(353) $3\frac{1}{2}$ (354) 5   (355) 3   (356) 8

(357) $2\frac{1}{7}$ (358) $6\frac{2}{5}$ (359) $3\frac{1}{3}$ (360) $1\frac{5}{8}$

FRACTIONS

Basic Math Answer Keys

(361) 7

(362) $3\dfrac{1}{2}$

(363) 2

(364) 5

(365) 1

(366) $\dfrac{5}{7}$

(367) 3

(368) $2\dfrac{1}{3}$

(369) $5\dfrac{1}{4}$

(370) 3

(371) $9\dfrac{1}{3}$

(372) $2\dfrac{3}{4}$

(373) 5

(74) $3\dfrac{1}{2}$

(375) 1

(376) $11\dfrac{7}{90}$

(377) $9\dfrac{17}{24}$

(378) $12\dfrac{5}{8}$

(379) $3\dfrac{89}{132}$

(380) $7\dfrac{69}{110}$

(381) $4\dfrac{13}{21}$

(382) $8\dfrac{5}{56}$

(383) 16

(384) $4\dfrac{11}{12}$

(385) $6\dfrac{17}{504}$

(386) $4\dfrac{4}{9}$

(387) $2\dfrac{23}{24}$

(388) $7\dfrac{271}{280}$

(389) $3\dfrac{2}{3}$

(390) $10\dfrac{23}{24}$

(391) $12\dfrac{37}{132}$

(392) $8\dfrac{5}{24}$

(393) $7\dfrac{47}{60}$

(394) $4\dfrac{2}{9}$

(395) 5

(396) $2\dfrac{29}{63}$

 **FRACTIONS**

Basic Math Answer Keys

(397) $6\frac{39}{56}$ (398) $1\frac{9}{20}$ (399) $2\frac{25}{33}$ (400) $7\frac{29}{35}$

(401) $12\frac{6}{7}$ (402) $6\frac{7}{11}$ (403) $14\frac{9}{13}$ (404) $21\frac{11}{13}$

(405) $6\frac{3}{14}$ (406) $9\frac{3}{5}$ (407) $9\frac{3}{5}$ (408) $5\frac{9}{11}$

(409) $4\frac{6}{7}$ (410) $16\frac{3}{5}$ (411) $11\frac{4}{7}$ (412) 15

(413) 10 (414) $17\frac{5}{9}$ (415) $22\frac{7}{13}$ (416) 16

(417) $9\frac{4}{7}$ (418) $7\frac{12}{13}$ (419) $18\frac{2}{7}$ (420) $15\frac{5}{13}$

(421) $18\frac{7}{8}$ (422) 11 (423) $12\frac{7}{16}$ (424) $13\frac{1}{2}$

(425) $27\frac{1}{11}$ (426) $2\frac{2}{15}$ (427) $3\frac{16}{35}$ (428) $\frac{47}{56}$

(429) $2\frac{1}{40}$ (430) $2\frac{7}{10}$ (431) $3\frac{19}{24}$ (432) $1\frac{19}{28}$

FRACTIONS

Basic Math Answer Keys

(433) $1\dfrac{1}{8}$ (434) $\dfrac{7}{8}$ (435) 4 (436) $1\dfrac{1}{5}$

(437) $7\dfrac{13}{42}$ (438) 2 (439) $4\dfrac{17}{21}$ (440) $2\dfrac{19}{28}$

(441) $3\dfrac{1}{15}$ (442) $2\dfrac{2}{7}$ (443) $7\dfrac{2}{3}$ (444) $6\dfrac{1}{14}$

(445) $4\dfrac{7}{12}$ (446) $1\dfrac{5}{8}$ (447) $5\dfrac{2}{3}$ (448) $6\dfrac{11}{12}$

(449) $2\dfrac{7}{8}$ (450) $1\dfrac{1}{3}$ (451) $8\dfrac{9}{10}$ (452) $22\dfrac{139}{210}$

(453) $9\dfrac{5}{6}$ (454) $17\dfrac{9}{112}$ (455) $15\dfrac{1}{48}$ (456) $4\dfrac{19}{42}$

(457) $5\dfrac{7}{48}$ (458) $2\dfrac{23}{56}$ (459) $3\dfrac{3}{10}$ (460) $6\dfrac{46}{315}$

(461) $3\dfrac{17}{30}$ (462) $3\dfrac{83}{120}$ (463) $2\dfrac{187}{210}$ (464) $8\dfrac{73}{77}$

(465) $9\dfrac{5}{6}$ (466) $3\dfrac{219}{440}$ (467) $1\dfrac{7}{9}$ (468) $17\dfrac{7}{20}$

FRACTIONS

Basic Math Answer Keys

(469) $6\dfrac{253}{312}$ (470) $20\dfrac{5}{6}$ (471) $13\dfrac{10}{11}$ (472) $16\dfrac{43}{120}$

(473) $14\dfrac{79}{154}$ (474) $9\dfrac{61}{112}$ (475) $9\dfrac{2}{3}$ (476) $6\dfrac{13}{210}$

(477) $3\dfrac{7}{30}$ (478) $5\dfrac{113}{420}$ (479) $3\dfrac{23}{24}$ (480) $6\dfrac{83}{168}$

(481) $5\dfrac{7}{24}$ (482) $7\dfrac{17}{20}$ (483) $2\dfrac{13}{14}$ (484) $6\dfrac{5}{6}$

(485) $7\dfrac{59}{84}$ (486) $10\dfrac{3}{8}$ (487) $15\dfrac{127}{140}$ (488) $9\dfrac{1}{30}$

(489) $11\dfrac{5}{12}$ (490) $4\dfrac{4}{21}$ (491) $9\dfrac{113}{140}$ (492) $11\dfrac{37}{42}$

(493) $7\dfrac{5}{24}$ (494) $14\dfrac{1}{15}$ (495) $3\dfrac{19}{24}$ (496) $8\dfrac{7}{8}$

(497) $10\dfrac{33}{140}$ (498) $7\dfrac{5}{6}$ (499) $8\dfrac{103}{140}$ (500) $6\dfrac{7}{12}$

(501) 1 (502) 2 (503) $\dfrac{2}{7}$ (504) 2

FRACTIONS

Basic Math Answer Keys

(505) $4\frac{1}{3}$ (506) $2\frac{1}{2}$ (507) 0 (508) 1

(509) 0 (510) $\frac{2}{3}$ (511) $\frac{6}{7}$ (512) 3

(513) 0 (514) $\frac{4}{5}$ (515) $\frac{1}{4}$ (516) $\frac{1}{2}$

(517) 0 (518) $3\frac{1}{2}$ (519) 1 (520) 1

(521) 1 (522) 0 (523) 2 (524) 1

(525) $2\frac{1}{8}$ (526) $\frac{4}{13}$ (527) $5\frac{7}{15}$ (528) $\frac{11}{16}$

(529) $6\frac{4}{5}$ (530) $6\frac{2}{5}$ (531) $5\frac{1}{12}$ (532) $\frac{1}{13}$

(533) $5\frac{1}{10}$ (534) $4\frac{1}{13}$ (535) $1\frac{5}{8}$ (536) $\frac{9}{14}$

(537) 3 (538) $13\frac{1}{2}$ (539) $4\frac{5}{9}$ (540) $2\frac{6}{7}$

FRACTIONS

Basic Math Answer Keys

(541) $\dfrac{1}{8}$ (542) $11\dfrac{2}{5}$ (543) $\dfrac{5}{7}$ (544) $3\dfrac{4}{7}$

(545) $6\dfrac{1}{2}$ (546) $\dfrac{7}{10}$ (547) $2\dfrac{9}{13}$ (548) $\dfrac{5}{9}$

(549) $4\dfrac{1}{9}$ (550) $3\dfrac{7}{13}$ (551) $\dfrac{4}{7}$ (552) $1\dfrac{19}{23}$

(553) $9\dfrac{11}{15}$ (554) $21\dfrac{10}{19}$ (555) $34\dfrac{27}{37}$ (556) $18\dfrac{22}{23}$

(557) $11\dfrac{12}{19}$ (558) $19\dfrac{7}{15}$ (559) $3\dfrac{37}{43}$ (560) 7

(561) $41\dfrac{9}{49}$ (562) 12 (563) $37\dfrac{24}{29}$ (564) $1\dfrac{2}{9}$

(565) $24\dfrac{3}{16}$ (566) $3\dfrac{2}{49}$ (567) $\dfrac{38}{41}$ (568) $1\dfrac{2}{5}$

(569) $3\dfrac{13}{23}$ (570) $\dfrac{32}{49}$ (571) $11\dfrac{1}{27}$ (572) $4\dfrac{16}{29}$

(573) $1\dfrac{26}{33}$ (574) $11\dfrac{9}{35}$ (575) $22\dfrac{13}{17}$ (576) $3\dfrac{11}{15}$

FRACTIONS

Basic Math Answer Keys

(577) $4\frac{7}{40}$ (578) $\frac{7}{8}$ (579) $\frac{1}{6}$ (580) 0

(581) $\frac{3}{35}$ (582) $\frac{5}{14}$ (583) $\frac{1}{3}$ (584) $2\frac{5}{24}$

(585) $1\frac{2}{35}$ (586) $2\frac{5}{8}$ (587) $1\frac{9}{35}$ (588) $\frac{7}{15}$

(589) $1\frac{3}{8}$ (590) $4\frac{1}{2}$ (591) $\frac{5}{8}$ (592) $1\frac{9}{28}$

(593) $3\frac{1}{15}$ (594) $1\frac{2}{3}$ (595) $\frac{1}{3}$ (596) $1\frac{17}{18}$

(597) $\frac{9}{20}$ (598) $4\frac{73}{120}$ (599) $4\frac{4}{9}$ (600) $6\frac{5}{12}$

(601) $5\frac{181}{264}$ (602) $2\frac{7}{12}$ (603) $\frac{1}{30}$ (604) $1\frac{7}{12}$

(605) $\frac{14}{33}$ (606) $1\frac{1}{20}$ (607) $2\frac{33}{56}$ (608) $\frac{71}{90}$

(609) $1\frac{7}{12}$ (610) $3\frac{3}{8}$ (611) $4\frac{2}{5}$ (612) $10\frac{8}{21}$

FRACTIONS

Basic Math Answer Keys

(613) $1\dfrac{5}{14}$ (614) $5\dfrac{1}{2}$ (615) $6\dfrac{61}{63}$ (616) $\dfrac{313}{390}$

(617) $2\dfrac{697}{819}$ (618) $4\dfrac{8}{13}$ (619) $2\dfrac{151}{420}$ (620) $2\dfrac{169}{180}$

(621) $3\dfrac{823}{1872}$ (622) $\dfrac{1}{4}$ (623) $4\dfrac{39}{140}$ (624) $\dfrac{2059}{4368}$

(625) $4\dfrac{7}{12}$ (626) $2\dfrac{470}{1001}$ (627) $3\dfrac{199}{240}$ (628) $1\dfrac{43}{390}$

(629) $\dfrac{23}{28}$ (630) $\dfrac{13}{35}$ (631) $\dfrac{2347}{3276}$ (632) $10\dfrac{65}{77}$

(633) $2\dfrac{631}{924}$ (634) $3\dfrac{673}{1680}$ (635) $4\dfrac{1753}{1872}$ (636) $9\dfrac{5}{6}$

(637) 0 (638) $4\dfrac{1}{3}$ (639) $3\dfrac{47}{56}$ (640) $4\dfrac{5}{6}$

(641) $2\dfrac{9}{40}$ (642) $5\dfrac{25}{42}$ (643) 0 (644) $1\dfrac{1}{14}$

(645) $5\dfrac{7}{40}$ (646) $8\dfrac{5}{12}$ (647) $6\dfrac{1}{3}$ (648) $3\dfrac{55}{56}$

 **FRACTIONS**

Basic Math Answer Keys

(649) $4\dfrac{219}{280}$ (650) $4\dfrac{13}{30}$ (651) $3\dfrac{3}{8}$ (652) $5\dfrac{19}{42}$

(653) $5\dfrac{3}{10}$ (654) $5\dfrac{3}{10}$ (655) $4\dfrac{7}{12}$ (656) $6\dfrac{43}{168}$

(657) 1   (658) $2\dfrac{11}{12}$ (659) $1\dfrac{1}{3}$ (660) $1\dfrac{19}{35}$

(661) $\dfrac{10}{21}$ (662) 5   (663) $4\dfrac{5}{21}$ (664) $2\dfrac{1}{4}$

(665) $3\dfrac{23}{168}$ (666) $7\dfrac{1}{2}$ (667) 6   (668) 2

(669) 5   (670) $3\dfrac{3}{5}$ (671) 1   (672) 4

(673) $5\dfrac{5}{7}$ (674) 4   (675) $12\dfrac{1}{2}$ (676) $4\dfrac{2}{3}$

(677) $5\dfrac{2}{3}$ (678) $4\dfrac{4}{5}$ (679) $10\dfrac{1}{2}$ (680) $9\dfrac{3}{4}$

(681) 4   (682) 3   (683) 8   (684) 5

FRACTIONS

Basic Math Answer Keys

(685) $6\frac{1}{2}$ (686) 5 (687) $5\frac{1}{7}$ (688) 7

(689) 5 (690) $7\frac{3}{7}$ (691) $18\frac{1}{6}$ (692) 23

(693) $20\frac{3}{11}$ (694) $15\frac{8}{11}$ (695) $16\frac{9}{16}$ (696) $15\frac{4}{5}$

(697) $17\frac{1}{12}$ (698) 21 (699) $9\frac{1}{2}$ (700) $13\frac{2}{5}$

(701) $13\frac{9}{14}$ (702) $15\frac{2}{3}$ (703) $20\frac{4}{9}$ (704) $13\frac{11}{12}$

(705) $13\frac{1}{5}$ (706) $16\frac{9}{10}$ (707) 9 (708) 28

(709) $17\frac{5}{13}$ (710) $11\frac{4}{15}$ (711) 25 (712) $17\frac{6}{13}$

(713) 19 (714) $16\frac{3}{10}$ (715) $9\frac{1}{2}$ (716) $3\frac{1}{24}$

(717) $4\frac{25}{28}$ (718) $4\frac{31}{42}$ (719) $3\frac{11}{24}$ (720) $3\frac{1}{6}$

FRACTIONS

Basic Math Answer Keys

(721) $9\dfrac{33}{40}$ (722) $5\dfrac{29}{40}$ (723) $11\dfrac{33}{56}$ (724) $6\dfrac{3}{4}$

(725) $1\dfrac{37}{56}$ (726) $3\dfrac{3}{5}$ (727) $4\dfrac{5}{8}$ (728) $5\dfrac{7}{8}$

(729) $5\dfrac{1}{4}$ (730) $2\dfrac{23}{40}$ (731) $8\dfrac{17}{20}$ (732) $5\dfrac{1}{40}$

(733) $6\dfrac{7}{8}$ (734) $3\dfrac{11}{12}$ (735) $7\dfrac{1}{3}$ (736) $7\dfrac{39}{40}$

(737) $8\dfrac{7}{8}$ (738) $6\dfrac{7}{40}$ (739) $2\dfrac{5}{7}$ (740) $9\dfrac{1}{6}$

(741) $31\dfrac{57}{88}$ (742) $16\dfrac{95}{168}$ (743) $18\dfrac{53}{60}$ (744) $20\dfrac{1}{30}$

(745) $17\dfrac{29}{33}$ (746) $15\dfrac{97}{198}$ (747) $11\dfrac{1}{36}$ (748) $20\dfrac{5}{12}$

(749) $16\dfrac{173}{420}$ (750) $7\dfrac{553}{990}$ (751) $20\dfrac{6}{7}$ (752) $19\dfrac{47}{126}$

(753) $18\dfrac{13}{18}$ (754) $12\dfrac{5}{9}$ (755) $9\dfrac{9}{20}$ (756) $15\dfrac{1}{15}$

FRACTIONS

Basic Math Answer Keys

(757) $19\dfrac{19}{60}$ (758) $19\dfrac{27}{40}$ (759) $8\dfrac{64}{105}$ (760) $26\dfrac{1}{20}$

(761) $9\dfrac{10}{11}$ (762) $8\dfrac{163}{198}$ (763) $8\dfrac{5}{24}$ (764) $15\dfrac{1}{4}$

(765) $16\dfrac{19}{20}$ (766) $4\dfrac{1}{3}$ (767) 3 (768) $3\dfrac{1}{4}$

(769) $1\dfrac{3}{4}$ (770) 3 (771) $3\dfrac{2}{3}$ (772) $5\dfrac{2}{5}$

(773) $1\dfrac{1}{4}$ (774) 3 (775) 0 (776) 3

(777) 1 (778) $2\dfrac{4}{7}$ (779) $2\dfrac{1}{3}$ (780) 0

(781) $1\dfrac{2}{3}$ (782) $3\dfrac{1}{3}$ (783) 0 (784) $2\dfrac{2}{3}$

(785) 0 (786) 3 (787) $3\dfrac{1}{2}$ (788) $1\dfrac{1}{2}$

(789) $4\dfrac{1}{7}$ (790) $2\dfrac{2}{3}$ (791) $\dfrac{1}{2}$ (792) 0

FRACTIONS

Basic Math Answer Keys

(793) $\dfrac{1}{3}$ (794) $\dfrac{1}{4}$ (795) $\dfrac{1}{4}$ (796) $1\dfrac{4}{7}$

(797) $4\dfrac{1}{2}$ (798) $3\dfrac{1}{3}$ (799) 3 (800) 3

(801) $\dfrac{1}{4}$ (802) $\dfrac{1}{8}$ (803) $\dfrac{1}{8}$ (804) $\dfrac{2}{7}$

(805) $1\dfrac{1}{2}$ (806) $1\dfrac{4}{7}$ (807) $2\dfrac{3}{8}$ (808) $1\dfrac{5}{6}$

(809) $1\dfrac{5}{8}$ (810) $1\dfrac{1}{8}$ (811) $1\dfrac{3}{7}$ (812) $1\dfrac{1}{2}$

(813) $\dfrac{1}{4}$ (814) $\dfrac{1}{4}$ (815) 0 (816) $2\dfrac{19}{28}$

(817) $1\dfrac{11}{15}$ (818) $3\dfrac{7}{24}$ (819) $\dfrac{5}{7}$ (820) $\dfrac{1}{7}$

(821) $4\dfrac{3}{10}$ (822) $3\dfrac{5}{24}$ (823) $1\dfrac{1}{3}$ (824) $\dfrac{13}{15}$

(825) $\dfrac{11}{12}$ (826) 1 (827) $2\dfrac{1}{20}$ (828) $3\dfrac{3}{28}$

FRACTIONS

Basic Math Answer Keys

(829) $\dfrac{5}{42}$ (830) $6\dfrac{2}{3}$ (831) $1\dfrac{2}{15}$ (832) $2\dfrac{3}{10}$

(833) $1\dfrac{1}{24}$ (834) $2\dfrac{19}{56}$ (835) $1\dfrac{13}{15}$ (836) 2

(837) 0 (838) $2\dfrac{2}{3}$ (839) $3\dfrac{3}{56}$ (840) $1\dfrac{1}{8}$

(841) $\dfrac{4}{21}$ (842) $2\dfrac{89}{105}$ (843) $\dfrac{37}{40}$ (844) $\dfrac{7}{12}$

(845) $1\dfrac{11}{24}$ (846) $\dfrac{23}{168}$ (847) $\dfrac{11}{14}$ (848) $4\dfrac{13}{40}$

(849) $1\dfrac{19}{60}$ (850) $\dfrac{11}{12}$ (851) $1\dfrac{2}{15}$ (852) $2\dfrac{11}{35}$

(853) $\dfrac{283}{420}$ (854) $\dfrac{43}{84}$ (855) $\dfrac{41}{60}$ (856) $\dfrac{2}{3}$

(857) $\dfrac{11}{168}$ (858) $\dfrac{25}{42}$ (859) $\dfrac{155}{168}$ (860) $3\dfrac{3}{4}$

(861) $\dfrac{33}{40}$ (862) $1\dfrac{2}{3}$ (863) $2\dfrac{139}{420}$ (864) $\dfrac{83}{168}$

FRACTIONS

Basic Math Answer Keys

(865) $1\frac{7}{12}$ (866) $\frac{19}{180}$ (867) $5\frac{107}{132}$ (868) $10\frac{7}{15}$

(869) $9\frac{17}{66}$ (870) $7\frac{419}{660}$ (871) $2\frac{269}{420}$ (872) $4\frac{1}{924}$

(873) $6\frac{67}{72}$ (874) $3\frac{403}{440}$ (875) $8\frac{23}{60}$ (876) $\frac{9}{10}$

(877) $8\frac{5}{12}$ (878) $1\frac{13}{45}$ (879) $9\frac{141}{220}$ (880) $16\frac{4}{33}$

(881) $11\frac{7}{8}$ (882) $12\frac{17}{24}$ (883) $6\frac{2}{3}$ (884) $16\frac{41}{55}$

(885) $12\frac{38}{45}$ (886) $1\frac{1}{24}$ (887) $\frac{3}{4}$ (888) $5\frac{601}{693}$

(889) $1\frac{2}{9}$ (890) $20\frac{43}{120}$ (891) $14\frac{1}{30}$ (892) $12\frac{19}{44}$

(893) $16\frac{309}{770}$ (894) $15\frac{65}{924}$ (895) $6\frac{9}{28}$ (896) $\frac{10}{21}$

(897) $4\frac{8}{9}$ (898) $\frac{7}{18}$ (899) $\frac{3}{8}$ (900) $6\frac{3}{7}$

FRACTIONS

Basic Math Answer Keys

(901) $2\frac{1}{3}$ (902) $2\frac{1}{4}$ (903) $3\frac{9}{40}$ (904) $1\frac{1}{20}$

(905) $\frac{1}{24}$ (906) $\frac{9}{10}$ (907) $1\frac{17}{18}$ (908) $\frac{5}{8}$

(909) $5\frac{1}{5}$ (910) $3\frac{1}{5}$ (911) $4\frac{11}{16}$ (912) $\frac{1}{36}$

(913) $13\frac{19}{32}$ (914) $\frac{3}{7}$ (915) $2\frac{14}{27}$ (916) $3\frac{3}{5}$

(917) $7\frac{23}{36}$ (918) $\frac{2}{5}$ (919) $\frac{81}{100}$ (920) $\frac{1}{4}$

(921) $\frac{57}{100}$ (922) $6\frac{20}{27}$ (923) $\frac{1}{21}$ (924) $122\frac{121}{144}$

(925) $80\frac{5}{24}$ (926) $\frac{8}{15}$ (927) $37\frac{1}{3}$ (928) $4\frac{37}{80}$

(929) $8\frac{1}{16}$ (930) $1\frac{7}{9}$ (931) $8\frac{104}{225}$ (932) $11\frac{1}{2}$

(933) $4\frac{1}{2}$ (934) $8\frac{379}{900}$ (935) $3\frac{95}{392}$ (936) $\frac{13}{180}$

FRACTIONS

Basic Math Answer Keys

(937) $3\dfrac{15}{28}$ (938) $46\dfrac{17}{28}$ (939) $24\dfrac{37}{240}$ (940) $33\dfrac{1}{108}$

(941) $1\dfrac{5}{24}$ (942) $7\dfrac{1}{32}$ (943) $15\dfrac{11}{27}$ (944) $\dfrac{52}{63}$

(945) 4 (946) $-22\dfrac{13}{990}$ (947) $-1\dfrac{19}{72}$ (948) $\dfrac{4}{5}$

(949) -40 (950) $-3\dfrac{63}{64}$ (951) 65 (952) $-95\dfrac{1}{3}$

(953) $-6\dfrac{55}{56}$ (954) $-597\dfrac{67}{99}$ (955) $-493\dfrac{23}{80}$ (956) $7\dfrac{1}{5}$

(957) $1\dfrac{1123}{4752}$ (958) $\dfrac{189}{320}$ (959) $-90\dfrac{40}{77}$ (960) $6\dfrac{3}{5}$

(961) $15\dfrac{625}{864}$ (962) $40\dfrac{17}{18}$ (963) $-25\dfrac{157}{450}$ (964) $1\dfrac{25}{32}$

(965) $7\dfrac{7}{9}$ (966) $-1\dfrac{1}{5}$ (967) $\dfrac{7}{10}$ (968) $\dfrac{15}{77}$

(969) $-2\dfrac{71}{80}$ (970) $41\dfrac{5}{12}$ (971) $26\dfrac{14}{81}$ (972) $1\dfrac{4}{27}$

FRACTIONS

Basic Math Answer Keys

(973) $14\frac{25}{36}$ (974) $8\frac{4}{5}$ (975) $5\frac{13}{48}$ (976) $10\frac{5}{16}$

(977) $14\frac{3}{35}$ (978) $22\frac{1}{7}$ (979) $22\frac{22}{35}$ (980) $9\frac{27}{32}$

(981) $12\frac{2}{7}$ (982) $6\frac{16}{45}$ (983) $1\frac{3}{4}$ (984) $1\frac{25}{32}$

(985) 24 (986) $7\frac{37}{56}$ (987) $28\frac{21}{50}$ (988) $3\frac{5}{6}$

(989) $24\frac{19}{24}$ (990) $2\frac{5}{14}$ (991) $24\frac{5}{14}$ (992) $13\frac{13}{24}$

(993) $10\frac{1}{16}$ (994) $\frac{7}{9}$ (995) $6\frac{17}{60}$ (996) $76\frac{2}{3}$

(997) $5\frac{115}{121}$ (998) $27\frac{61}{132}$ (999) $48\frac{117}{140}$ (1000) $5\frac{17}{605}$

(1001) $15\frac{17}{33}$ (1002) $25\frac{239}{720}$ (1003) $9\frac{3}{4}$ (1004) $\frac{221}{240}$

(1005) $27\frac{421}{512}$ (1006) $2\frac{569}{720}$ (1007) $105\frac{5}{7}$ (1008) $67\frac{43}{56}$

FRACTIONS

Basic Math Answer Keys

(1009) $5\frac{10}{11}$ (1010) $272\frac{5}{14}$ (1011) $4\frac{27}{112}$ (1012) $2\frac{19}{20}$

(1013) $58\frac{37}{48}$ (1014) $8\frac{1129}{1188}$ (1015) $28\frac{7}{132}$ (1016) $86\frac{1}{4}$

(1017) $89\frac{109}{144}$ (1018) $34\frac{221}{660}$ (1019) $227\frac{20}{33}$ (1020) $1\frac{139}{176}$

(1021) $\frac{1}{6}$ (1022) $\frac{20}{21}$ (1023) $\frac{81}{25}$ (1024) $\frac{9}{4}$

(1025) $\frac{2}{9}$ (1026) $\frac{1}{3}$ (1027) $\frac{14}{3}$ (1028) 3

(1029) 0 (1030) $\frac{13}{12}$ (1031) $\frac{1}{2}$ (1032) $\frac{32}{3}$

(1033) $\frac{85}{36}$ (1034) $\frac{195}{56}$ (1035) $\frac{9}{8}$ (1036) $\frac{7}{4}$

(1037) $\frac{13}{4}$ (1038) $\frac{3}{4}$ (1039) $\frac{56}{45}$ (1040) $\frac{10}{21}$

(1041) $\frac{4}{3}$ (1042) $\frac{2}{3}$ (1043) $\frac{27}{16}$ (1044) $\frac{57}{10}$

FRACTIONS

Basic Math Answer Keys

(1045) $\dfrac{52}{21}$ (1046) $\dfrac{65}{98}$ (1047) $\dfrac{5}{3}$ (1048) $\dfrac{16}{21}$

(1049) $\dfrac{57}{10}$ (1050) $\dfrac{1}{3}$ (1051) $\dfrac{96}{35}$ (1052) $\dfrac{11}{45}$

(1053) $\dfrac{459}{100}$ (1054) $\dfrac{9}{4}$ (1055) $\dfrac{130}{63}$ (1056) $\dfrac{5}{3}$

(1057) $\dfrac{144}{25}$ (1058) $\dfrac{64}{21}$ (1059) $\dfrac{2}{3}$ (1060) 0

(1061) $\dfrac{15}{8}$ (1062) $\dfrac{5}{21}$ (1063) $\dfrac{5}{16}$ (1064) $\dfrac{110}{21}$

(1065) $\dfrac{77}{3}$ (1066) $\dfrac{18}{5}$ (1067) $\dfrac{10}{9}$ (1068) $\dfrac{40}{49}$

(1069) $\dfrac{3}{7}$ (1070) $\dfrac{297}{140}$ (1071) $\dfrac{3}{50}$ (1072) $\dfrac{156}{161}$

(1073) $12\dfrac{1}{2}$ (1074) 12 (1075) $\dfrac{1}{6}$ (1076) $4\dfrac{3}{8}$

(1077) $3\dfrac{1}{21}$ (1078) $1\dfrac{14}{19}$ (1079) $\dfrac{7}{17}$ (1080) $2\dfrac{4}{5}$

FRACTIONS

Basic Math Answer Keys

(1081) $1\frac{1}{3}$ (1082) $1\frac{1}{14}$ (1083) $1\frac{1}{9}$ (1084) $1\frac{1}{6}$

(1085) $1\frac{4}{5}$ (1086) 33 (1087) $2\frac{1}{4}$ (1088) $\frac{12}{13}$

(1089) $5\frac{3}{4}$ (1090) $\frac{7}{8}$ (1091) $\frac{3}{4}$ (1092) $\frac{1}{11}$

(1093) $1\frac{3}{10}$ (1094) $\frac{7}{9}$ (1095) $8\frac{1}{10}$ (1096) $\frac{8}{9}$

(1097) $9\frac{1}{7}$ (1098) $16\frac{1}{9}$ (1099) $\frac{21}{25}$ (1100) $2\frac{2}{5}$

(1101) $\frac{136}{301}$ (1102) $1\frac{1}{26}$ (1103) $\frac{115}{147}$ (1104) $1\frac{9}{43}$

(1105) $1\frac{4}{21}$ (1106) $1\frac{13}{15}$ (1107) $\frac{99}{130}$ (1108) $1\frac{1}{68}$

(1109) $\frac{9}{32}$ (1110) 1 (1111) $2\frac{29}{35}$ 1(112) $\frac{5}{9}$

(1113) $\frac{82}{99}$ (1114) $1\frac{13}{68}$ (1115) $5\frac{15}{16}$ (1116) $1\frac{43}{56}$

FRACTIONS

Basic Math Answer Keys

(1117) $\dfrac{15}{44}$ (1118) $1\dfrac{1}{11}$ (1119) $3\dfrac{19}{22}$ (1120) $\dfrac{40}{51}$

(1121) $\dfrac{1}{3}$ (1122) 4 (1123) $\dfrac{21}{22}$ (1124) $\dfrac{25}{18}$

(1125) $\dfrac{4}{3}$ (1126) 6 (1127) $\dfrac{7}{20}$ (1128) 3

(1129) $\dfrac{9}{4}$ (1130) $\dfrac{2}{3}$ (1131) $\dfrac{3}{2}$ (1132) $\dfrac{99}{64}$

(1133) $\dfrac{19}{20}$ (1134) 1 (1135) 1 (1136) $\dfrac{5}{12}$

(1137) $\dfrac{81}{98}$ (1138) $\dfrac{8}{45}$ (1139) 1 (1140) $\dfrac{2}{3}$

(1141) $\dfrac{3}{2}$ (1142) $\dfrac{2}{15}$ (1143) $\dfrac{32}{55}$ (1144) 9

(1145) $\dfrac{1}{4}$